Mapper of Mountains

✳

M.P. BRIDGLAND
in the CANADIAN ROCKIES
1902–1930

MAPPER OF MOUNTAINS

✳

M.P. BRIDGLAND

in the CANADIAN ROCKIES

1902–1930

———

I.S. MacLAREN

{ WITH ERIC HIGGS
AND GABRIELLE
ZEZULKA-MAILLOUX

 THE UNIVERSITY
OF ALBERTA PRESS

Published by
The University of Alberta Press
Ring House 2
Edmonton, Alberta, Canada T6G 2E1

MOUNTAIN CAIRNS ▲ *A series on the history and culture of the Canadian Rocky Mountains.*

ISBN 0–88864–456–6

Library and Archives Canada Cataloguing in Publication

MacLaren, I. S., 1951–
    Mapper of mountains : M.P. Bridgland in the Canadian Rockies
1902-1930 / I.S. MacLaren ; with Eric Higgs and Gabrielle Zezulka-Mailloux.

Includes bibliographical references and index.
ISBN 0-88864-456-6

    1. Bridgland, M. P. (Morrison Parsons), 1878–1948.   2. Mountain
mapping—Alberta—Jasper National Park—History—20th century.
3. Photographic surveying—Alberta—Jasper National Park—History—20th
century. 4. Cartographers—Canada—Biography.   5. Surveyors—Canada—Biography.
I. Higgs, Eric S.   II. Zezulka-Mailloux, Gabrielle E. M. (Gabrielle Eva Marie), 1974–
III. Title.

GA473.7.B74M32 2005   526'.092   C2005–906918–X

The University of Alberta Press is committed to protecting our natural environment. As part of our efforts,
this book is printed on stock produced by New Leaf Paper: it contains 100% post-consumer recycled fibres and
is acid- and chlorine-free.

The University of Alberta Press gratefully acknowledges the support received for its publishing program from The
Canada Council for the Arts. The University of Alberta Press also gratefully acknowledges the financial support of the
Government of Canada through the Book Publishing Industry Development Program (BPDIP) and from the Alberta
Foundation for the Arts for our publishing activities.

CONTENTS

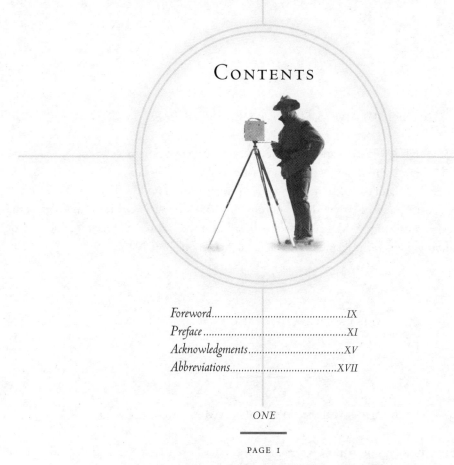

Foreword...IX

Preface...XI

Acknowledgments..................................XV

Abbreviations.......................................XVII

ONE

PAGE 1

MEASURING THE WEST BEFORE BRIDGLAND
EXPLORING, SURVEYING, AND MAPPING

Early Explorers, Surveyors, and Map-makers ✳ The Advent of the Dominion Lands Survey ✳ Railway Survey in the Athabasca Valley, 1872 ✳ Surveyors Strive to Stay Ahead of the Railway Builders ✳ Surveying the Canadian Rocky Mountains Using Phototopography ✳ Notes

TWO

PAGE 45

BRIDGLAND'S LIFE AND TIMES 1878–1914

Bridgland's Roots ✳ Bridgland's Early Years ✳ A Career of Adventure ✳ Bridgland's First Year in the Mountains ✳ Rocks and Hard Places ✳ Surveying the Main Range of the Canadian Rockies ✳ Marriage and Family ✳ The Alpine Club of Canada Is Born ✳ The ACC's First Annual Summer Camp, 1906 ✳ ACC Summer Camps, 1907–1911 ✳ Bridgland the Consummate Guide ✳ Triangulation Surveys of the Railway Belt, Crowsnest Forest Reserve, and Waterton Lakes Park ✳ Notes

THREE

PAGE 99

BRIDGLAND'S SURVEY OF JASPER PARK 1915

One Last Photograph ✽ Creating a Park ✽ More Than a Normal Summer's Work ✽ First Peak of the Season—Signal Mountain ✽ Bridgland's Photographic Methods ✽ More Peaks and Valleys ✽ Inside Bridgland's Dark-tent ✽ Surveying to the South and North ✽ A Day Off, A Life Lost ✽ August Along the Athabasca ✽ Autumn in the Valley ✽ Once Home in Calgary, the Challenge of Map-making ✽ Mapping from the Phototopographic Surveys ✽ The Art of Bridgland's Photographs ✽ Bridgland's Monumental Achievement ✽ Notes

PAGE 159

M.P. BRIDGLAND'S MAPS OF JASPER

FOUR

PAGE 169

BRIDGLAND'S LIFE AND TIMES 1916–1948

M.P. Becomes an Author, 1916 ✽ Bow River and Clearwater Forest Reserves, and Beyond, All the Way to Pitt Lake, 1917–1921 ✽ Surveying Kootenay Park, 1922–1923 ✽ Surveys Engineer, 1924–1931 ✽ Forced Retirement ✽ Bridgland's Later Life, 1931–1948 ✽ Mapper of Mountains ✽ Notes

FIVE

PAGE 211

BRIDGLAND'S LEGACY
*CULTURE, ECOLOGY, AND RESTORATION IN JASPER NATIONAL PARK,
AND THE ROCKY MOUNTAIN REPEAT PHOTOGRAPHY PROJECT*

Tracking Ecological and Cultural Changes by Photographs ✽ The Rocky Mountain Repeat Photography Project ✽ Comparing the Original and Repeat Photographs of Jasper ✽ Battling the Scythe of Old Man Time—Preserving Both Sets of Photographs ✽ Bridgland's Legacy ✽ Notes

PAGE 243

ROCKY MOUNTAIN REPEAT PHOTOGRAPHY PROJECT

Appendices...............................265
Sources...................................271
Index.....................................283

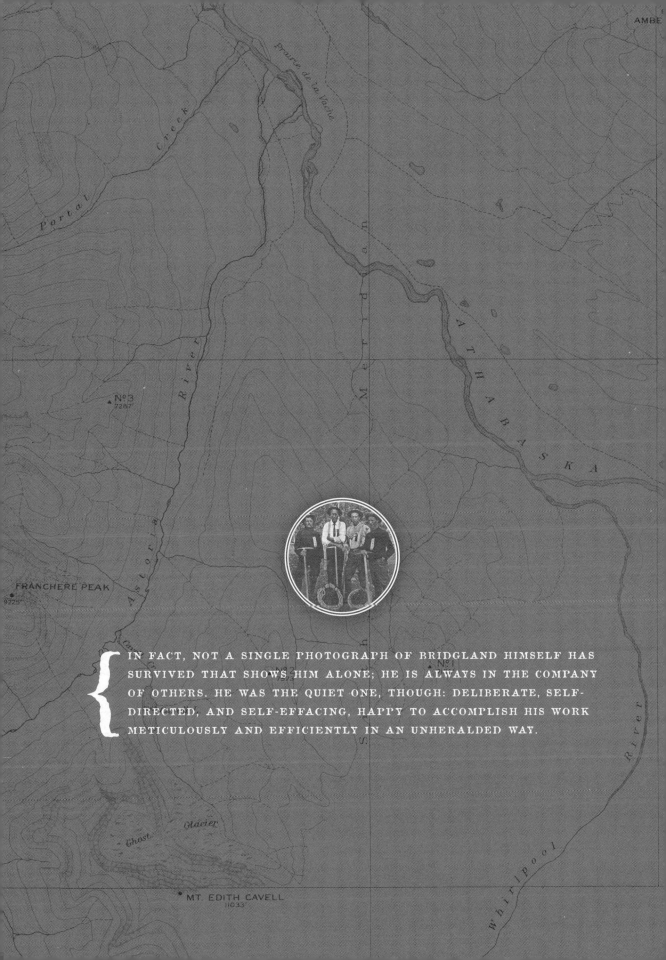

IN FACT, NOT A SINGLE PHOTOGRAPH OF BRIDGLAND HIMSELF HAS
SURVIVED THAT SHOWS HIM ALONE; HE IS ALWAYS IN THE COMPANY
OF OTHERS. HE WAS THE QUIET ONE, THOUGH: DELIBERATE, SELF-
DIRECTED, AND SELF-EFFACING, HAPPY TO ACCOMPLISH HIS WORK
METICULOUSLY AND EFFICIENTLY IN AN UNHERALDED WAY.

Photograph by M.P. Bridgland of Mt Bridgland from Mt Clairvaux, Stn 18, no 159, direction northwest, 1915.
Mt Bridgland is the prominent peak in the centre of the three in the middle of the view, looking up the Miette
River valley. It went by the name of Toot Toot until it was officially named in 1923. Courtesy Jasper National Park.

FOREWORD

CHILDHOOD MEMORIES of Calgary's dark, long, winter evenings contrast with happy recollections of Uncle Morris's invitations to his home to share in the excitement of his summer adventures through his meticulously made photographs and bright, hand-coloured photographic slides. These pictures allowed his family and friends to participate in his experience of surveying mountain peaks and valleys previously documented only by earlier map-makers such as David Thompson.

While he was an undergraduate student at the University of Toronto, his interest in photography developed. Graduating to a Dominion Lands Surveyor, Morris elaborated his interest and perfected his photographic technique in mountain surveying. From the start of the twentieth century, Dominion Lands Surveyor Bridgland annually and patiently climbed the Canadian Rocky Mountains to photograph and use his skills as a civil engineer to make the maps that gave Canadians of his era their knowledge of this great region of the nation and nourish their appetite to explore the country and commence developing its natural resources. Healthy, invigorating activities of mountain climbing and holidaying were facilitated by his maps and the energy with which he helped establish the Alpine Club of Canada and run its first summer camps in the mountains.

Large, gentle, and soft-spoken, Morris was friendly and generous with his time, knowledge, and experiences. Despite spending long sedentary winters producing the tracing sheets for his maps and painting his survey slides, Morris conscientiously kept in physical condition to be prepared for each successive summer's strenuous fieldwork. Whether as chief of party with his Dominion Lands Survey crews or as head guide and chief mountaineer for the Alpine Club of

BRIDGLAND ANNUALLY AND PATIENTLY CLIMBED THE CANADIAN ROCKY MOUNTAINS TO PHOTOGRAPH AND USE HIS SKILLS AS A CIVIL ENGINEER TO MAKE THE MAPS THAT GAVE CANADIANS OF HIS ERA THEIR KNOWLEDGE OF THIS GREAT REGION OF THE NATION.

Canada, he patiently developed the confidence, interests, and abilities of children, young survey assistants, and amateur alpinists. This mentorship was recognized by his fellow professionals, alpinists, relations, and friends.

With the required perseverance, if upon reaching a peak, were he to see a still higher mountain, he would ascend it and take another photograph. This was a lesson not lost on me when I thought I had something accomplished and likewise found it incomplete. Uncle Morris was thorough and tireless in those tasks, hobbies, and people to which and to whom he devoted himself.

After he retired from surveying, he would take his son Edgar (now Brigadier General Bridgland of Canadian Aviation Development fame) and me into the mountains for our summer holiday. He shared his skills with us, fulfilling our desire to be part of our hero's adventure. Just as Morris developed techniques to gain topographic knowledge, so did his son, through aviation.

Well I recall telling friends at school of the views we saw from his unusual lookouts. When high school trigonometry seemed my nemesis, he spent precious evenings coaching me and translating lines and angles into mountains until I came to appreciate one of trigonometry's practical applications and values.

At the University of Toronto, our shared *alma mater*, he persuaded me to attend the first classes of Geography in Canada given by Dr. Griffith Taylor. Indeed, my lifetime experiences always have moments and memories of this wonderful uncle Morris and his tolerance and belief in expanding horizons for his only little niece. In the summer of 2003, while attending the Jasper-Yellowhead Historical Museum and Archives' exhibit of M.P. Bridgland's Jasper survey photographs of 1915, I was struck and gratified to see that the application to ecological research continues to broaden all our horizons.

Thank you Ian MacLaren, Eric Higgs, and Gabrielle Zezulka-Mailloux, for your knowledge and understanding of this man, that we may, through reading this thoroughly researched book, *Mapper of Mountains*, come to know this wonderful, life-broadening, and sharing man, my uncle Morris Bridgland. ✳

I WAS STRUCK AND GRATIFIED TO SEE THAT THE APPLICATION TO ECOLOGICAL RESEARCH CONTINUES TO BROADEN ALL OUR HORIZONS.

MARGARET HESS
Calgary, Alberta

Preface

WHEN YOU THINK OF SURVEYING, have you an image of a solitary person exposed to the elements in some remote part of the country, back to the wind and shivering in winter, behatted, dusty, and mosquito bitten in summer? Fur-trade-era giant David Thompson, unsung as the pre-eminent early mapper of the North American West, leaps to mind as the type. Similarly unsung is Dominion Lands Surveyor and Alpine Club of Canada co-founder Morrison Parsons Bridgland (1878–1948). In 1915, just over a century after Thompson and his brigade made their way up the Athabasca and Whirlpool rivers to Athabasca Pass early in 1811 and then descended into the Columbia River watershed on his great voyage of discovery down that river to the sea, Bridgland, our Mapper of Mountains, brought his unmatched skills as a surveyor to the same valley.

We were not left with a strong understanding of what Jasper National Park looked like to David Thompson two centuries ago. What did it look like a century ago? Bridgland's maps were the ones that early tourists used to explore the natural wonders of this part of the eastern Rockies, and his book, *Description of & Guide to Jasper Park* (1917) told them what to go and see. He climbed many of the park's mountains for the first time. Yet he was seldom alone. Like Thompson he travelled in groups. He worked with groups of men as he conducted the phototopographic and transit surveys that he would use to produce those maps. In fact, not a single photograph of Bridgland himself has survived that shows him alone; he is always in the company of others. He was the quiet one, though: deliberate, self-directed, and self-effacing, happy to accomplish his work meticulously and efficiently in an unheralded way. In his era, Arthur Oliver Wheeler sought and gained the fame that was apparently inconsequential to the Mapper of

IN 1915, JUST OVER A CENTURY AFTER THOMPSON AND HIS BRIGADE MADE THEIR WAY UP THE ATHABASCA AND WHIRLPOOL RIVERS TO ATHABASCA PASS EARLY IN 1811 AND THEN DESCENDED INTO THE COLUMBIA RIVER WATERSHED ON HIS GREAT VOYAGE OF DISCOVERY DOWN THAT RIVER TO THE SEA, BRIDGLAND, OUR MAPPER OF MOUNTAINS, BROUGHT HIS UNMATCHED SKILLS AS A SURVEYOR TO THE SAME VALLEY.

Mountains. Whether working for the federal government's Dominion Lands Survey or scaling peaks with the Alpine Club of Canada, M.P. Bridgland contributed so extensively to the history of this part of Canada that you will find yourself surprised that his name is not as familiar to you as the names of the most famous surveyors, cartographers, alpinists, and peaks of North America. From the age of twenty-three to the age of fifty-one, Bridgland spent nearly every summer mapping parts of the mountains of Alberta or British Columbia.

Mapper of Mountains takes its reader through a brief history of surveying in western Canada in order to provide the context for Bridgland's life, times, and career of mountain surveying, which extended from 1902 to 1930. It explains how photography became a staple in mountain surveying. Bridgland was charged with making maps, but he needed to conduct both phototopographic and triangulation surveys in the mountains to gather the necessary data before the work of cartography could occur back in the Calgary office. Next, the book highlights a summer survey ninety years ago of the principal sites of Jasper Park that were attracting tourists in the first decade of its existence. (Its name was changed to Jasper National Park in 1930.) How Bridgland made his photographs from the tops of mountains and even developed them while camped out in the wilderness are detailed, as are some of the trials and tribulations involved in that summer's survey. *Mapper of Mountains* also relates the personal history of Bridgland and his family, the work of the other years of his career, and his involvement in the establishment and early years of the Alpine Club of Canada.

IN 1997 A TEAM OF ACADEMICS RESEARCHING THE ECOLOGICAL HISTORY OF JASPER NATIONAL PARK BEGAN A SYSTEMATIC PROJECT OF REPHOTOGRAPHING THE PARK FROM ALL NINETY-TWO OF THE MOUNTAIN PEAK STATIONS THAT BRIDGLAND CHOSE FOR HIS PHOTOGRAPHS SO LONG AGO AND STUDYING THE FINDINGS.

In 1997 a team of academics researching the ecological history of Jasper National Park began a systematic project of rephotographing the park from all ninety-two of the mountain peak stations that Bridgland chose for his photographs so long ago and studying the findings. Now there are two sets of photographs shot more than eighty years apart. Together, they permit comparative studies of changes in vegetation and human influence on non-human nature, as well as the effects on the landscape imposed by park staff since it fell under park designation in 1911. The rephotographers' story is told as part of Bridgland's monumental legacy.

That legacy is enduring and full of vitality for park visitor and researcher alike. Emerging out of his surveyor's and our researcher's milieux into a public light, Bridgland deserves to have his story told widely in word and image. Because the rephotography of Jasper led to a second project, rephotographing Waterton Lakes National Park, in order to provide a comparable set of photos to the ones made by Bridgland and his teams over the course of the two summers—1913 and 1914—prior to the Jasper survey, the legacy is growing and expanding. Now known as the Rocky Mountains Repeat Photography Project (http://bridgland.sunsite.ualberta.ca/), this second phase of research based on Bridgland's surveys cried out for a scouring of the historical record to find what it holds about the man himself.

The task would not be an easy one: M.P. was a man of few words outside of his job. There are no personal letters extant, and few others wrote about him, probably because he made no effort to attract attention to himself. So locating and making contact with Edgar, Bridgland's second and only surviving son, was a key to putting M.P.'s life together. Making the acquaintance of Margaret B. Hess, who knew M.P. well during his Calgary years, proved another key to the biographical component of the story.

Next, there was the matter of trying to locate all his Jasper-area and Waterton-area work. After all, this was a man whose identity *was* his work; a biography of him was going to depend heavily on the output of his career. Sleuthing both in provincial and national repositories turned up nothing initially, but in Ottawa Edgar was certain that as recently as the early 1990s he had seen at least the glass plate negatives made in the field by M.P. and his assistants. The Geodetic Survey of Canada, the Legal Surveys Division of Natural Resources Canada, and Library and Archives Canada (formerly the National Archives of Canada) were contacted, but they all confirmed that the paper trail about the plates came to an inconclusive halt. A few years of fruitless wondering, poking about, and emailing ensued. Then, with the aid of an archivist, a building used for storage by the Archives down on Tunney's Pasture in Ottawa yielded its treasure. There, in cardboard boxes, many of the glass plate negatives rewarded perseverance. They languished uncatalogued and forgotten by overworked achivists, yes, but not lost! The Jasper year's plates were there and so were those for other years, including the previous two, the Waterton years.

The photographs made from these sets of glass plates and the plates themselves formed the basis for understanding the story of M.P.'s twenty-five or so summers of work well enough to tell it here. Fitfully, his field diaries and other written sources turned up, as well.

NOW KNOWN AS THE ROCKY MOUNTAINS REPEAT PHOTOGRAPHY PROJECT, THIS SECOND PHASE OF RESEARCH BASED ON BRIDGLAND'S SURVEYS CRIED OUT FOR A SCOURING OF THE HISTORICAL RECORD TO FIND WHAT IT HOLDS ABOUT THE MAN HIMSELF.

Long after they could have helped the rephotographers in Jasper, so did the transit notes containing the latitude and longitude readings for the stations from which the Mapper of Mountains and his crews shot their photographs. Edgar was as thrilled as we at the chance to bring the story to light.

THE FOLLOWING PAGES
AIM, THEN, TO PUT
INTO WORD AND IMAGE
THE LEGACY THAT
BRIDGLAND'S
QUARTER-CENTURY OF
WORK HAS LEFT TO
EVERYONE INTERESTED
IN THE MOUNTAINS OF
THE CANADIAN WEST.

At presentations of our work, we learned about the keen interest that mountain surveying aroused in the general public, so we grew convinced that a book such as *Mapper of Mountains* needed to be written for the general, not the academic, reader. In particular, the annual general meeting in 2004 of the Alberta Land Surveyors' Association had made clear that the tradition of survey techniques was beginning to disappear from the collective memory of present-day surveyors, who voiced genuine appreciation over hearing the tale of M.P. and his survey methods, long since displaced by computerized field instruments. The following pages aim, then, to put into word and image the legacy that Bridgland's quarter-century of work has left to everyone interested in the mountains of the Canadian West. As old as the province of Alberta and the Alpine Club of Canada, his career deserves this attention particularly at the time of the province's and the club's centennials. ✳

ACKNOWLEDGEMENTS

※

As is all the work of the Rocky Mountain Repeat Photography Project (RMRPP), this book is supported by a research grant from the Social Sciences and Humanities Research Council of Canada. Publication grants were generously provided by anonymous donors who encourage and endorse research into western Canadian history, by the Alberta Land Surveyors' Association, by the Alberta Historical Resources Foundation's Publications Program, and by Foothills Model Forest, Hinton, Alberta. The authors express their deep thanks for this bountiful support.

Institutional support and encouragement was gratefully received from the following libraries, museums, archives, and associations: the Alberta Land Surveyors' Association, the Alpine Club of Canada, the American Alpine Club, the Archives of Ontario, the British Columbia Land Surveyors, the British Columbia Archives and Records Service, the Calgary Public Library, the E.J. Pratt Library at Victoria University (Toronto), the Earth Sciences Information Centre of Natural Resources Canada, the Glenbow Museum and Archives, the Jasper-Yellowhead Museum and Archives, the Library and Archives Canada, the Provincial Archives of Alberta, the Legal Surveys Division of Natural Resources Canada, the Thomas Fisher Rare Books Library at University of Toronto, the United Church/Victoria University (Toronto) Archives, the Archives of the Whyte Museum of the Canadian Rockies, and the University of Alberta's Bruce Peel Special Collections, Humanities and Social Sciences, and Cameron libraries.

For assistance in the contribution of specific information for this book, we heartily thank Harold Averill, Trish Bailey, Cheryl Baxter, Dave Birrell, Don Bourdon, David Boutillier and the Boutillier family, Jim Bowman, Sandy Campbell, Glenda Cornforth, Dave Cruden, Janet

Davis, Jill Delaney, Diana Fancher, Kim Forster, Bonnie Gallinger, Anne Goddard, Edna Hajnal, Bob Hallam, Curt Hegel, Greg Horne, Fred Howlett, Judy Larmour, Josee Larochelle, Deborah Lister, Fran Loft, Brian Luckman, Michael MacDonald, Eli MacLaren, Alec MacLeod, Eva Major-Marothy, Brian Munday, Peter Murphy, Ruth Oltmann, Art Peterson, Tom Peterson, Jeanine Rhemtulla, Zac Robinson, Dave Rothwell, Adrienne Shaw, Jim Simpson, J.S. Simpson, Trudi Smith, Apollonia Steele, Bob Stevenson, Lise Tataryn, Brenda Taylor, Jim Taylor, Amy Tourond, Kevin Van Tighem, Ota Vyskocil, Dwain Wacko, Rod Wallace, Rob Watt, Graham Watt-Gremm, Jenaya Webb, and Cliff White. Thanks as well go to the Alberta Land Surveyors' Association for hosting a presentation about Bridgland to delegates attending its annual general meeting in Jasper, April 2004.

Research assistant Gabrielle Zezulka-Mailloux undertook initial research, and all members of the RMRPP team extend their thanks to her for it. For her part, she acknowledges the input of her father, Joseph Zezulka.

Preparation of this work owes an inexpressible thanks to many people, including the staff of the University of Alberta Press, as well as other staff at the University of Alberta, Dietlind Bechthold, Linda Bridges, and Darren Shaw, in particular. Thanks also to the Press's free-lancers, Eva Radford who copyedited the manuscript, Wendy Johnson who drew the modern maps, and Lara Minja of Lime Design, who designed the book's interior and cover.

Special heartfelt thanks are extended to Edgar Bridgland and Margaret (Marmie) Hess for their information and keen interest in seeing this work published. ✳

Abbreviations

✳

ACC   Alpine Club of Canada

ALSA   Alberta Land Surveyors' Association

BCARS   British Columbia Archives and Records Service

CER   Culture, Ecology, and Restoration Research Project

CNoR   Canadian Northern Railway

CPR   Canadian Pacific Railway

DLS   Dominion Lands Survey /
               Dominion Lands Surveyor

FMF   Foothills Model Forest

GSC   Geodetic Survey of Canada

GTPR   Grand Trunk Pacific Railway

HBC   Hudson's Bay Company

HBCA, PAM   Hudson's Bay Company Archives,
               Provincial Archives of Manitoba

LAC   Library and Archives Canada
               (formerly, National Archives of Canada)

NWC   North West Company

RMRPP   Rocky Mountain Repeat Photography Project

{ THE CANADIAN GOVERNMENT PLANNED TO CAPITALIZE ON ITS NEWLY ACQUIRED LANDS AND NATURAL RESOURCES AS IT PROMOTED BOTH WESTERN SETTLEMENT AND THE BUILDING OF A RAILWAY.

MEASURING THE WEST
BEFORE BRIDGLAND

EXPLORING, SURVEYING, AND MAPPING

✳

Early Explorers, Surveyors, and Map-makers

What was the tradition of surveying and mapping the Canadian West that awaited Morrison Parsons Bridgland, fresh out of an undergraduate degree and a single term at the School of Practical Sciences, University of Toronto, when he came west for the first time early in 1902? And, specifically, what surveying had occurred in the upper Athabasca River valley prior to his prodigious summer of work there in 1915, at the middle of his illustrious career of mountain phototopographical surveying and map-making? The answers to these questions take up the first part of our story. They will provide readers of Bridgland's career with the necessary context to appreciate it.

If we trace the early history of western Canada, we find that the traditions of mapping elaborated and extended by Bridgland find their origins in Native North Americans and, thereafter, every European culture that explored and settled the continent. The concept of maps was known to Native Americans long before Europeans came to the New World, but these were not distinguished from other forms of oral or written records until after the Conquest by Spain. In 1519, painters in the court of Moctezuma supplied conquistador Hernán Cortés with a cloth map that enabled him to find safe harbour for his ships.[1] But in 1540, the Spanish explorer Hernando de Alarcón asked a Native American to "set [him] downe in a charte as much as he knew concerning [the Rio Colorado], and what manner of people those were which dwelt upon the banckes thereof on both sides: which he did...with pleasure...on a piece of paper."[2] Later that year, much farther north, four Iroquois used "little stickes, which they layde upon the ground in a certaine distance [representing the River(s)] and afterward layde other small branches between both, representing the Saults," to show Jacques Cartier, then on his third voyage of exploration in the St Lawrence River, the location of the Lachine Rapids upriver from the site that later became Montreal.[3] These were the beginnings of a long history of collaboration between Native American map-makers and European explorers, cartographers, and surveyors. European methods of surveying and map making eventually overtook Native ones, but even though they may have been more systematic, there is little evidence that they were more accurate initially. European methods, having long relied on cleared lands and human landmarks, met their match "in a country almost 'every where covered with woods. ... Here,'" wrote Lewis Evans as late as 1755 in the analysis accompanying his *General Map of the Middle British Colonies of America*, "'are no churches, Towers, Houses or peaked Mountains to be seen from afar, no Means of obtaining the Bearings or Distances of Places, but by the Compass, and actual Mensuration with the Chain.'"[4]

THE CONCEPT OF
MAPS WAS KNOWN TO
NATIVE AMERICANS
LONG BEFORE
EUROPEANS CAME TO
THE NEW WORLD.

As the population in New France increased and a demand for land maps arose, chiefly for the purposes of establishing property rights, settled areas and areas deemed important for trade or military purposes were charted by engineers and surveyors under the direction of military authorities. However, the interior lands, especially those west of Lake Superior, remained unmapped or only sketched. Although Henry Hudson, Robert Bylot, the Dane Jens Munk, Luke Foxe, and Thomas James had all surveyed coastal areas in Hudson Bay from their ships in the early seventeenth century, once the Hudson's Bay Company (HBC) came into being later in the century it commissioned few surveys. Before the fall of Quebec and the end of New France in 1759, Pierre Gaultier de Varennes et de la Vérendrye and his sons explored the interior as far as the Rockies, somewhere south of 49°N Lat. The HBC's lack of penetration into the interior had attracted the notice and publicly vented ire of Arthur Dobbs. Later the governor of North Carolina, Dobbs was appointed engineer-in-chief and surveyor general in Ireland in 1730. A man of wealth and liberal views, he was keen on Britain's discovery of a Northwest Passage by which he could facilitate trade between Britain and both India and China. His criticism spurred mid-eighteenth-century voyages, but these all succeeded only in confirming that Hudson Bay formed a western extension of the Atlantic Ocean, not a passage out of it to another body of water.

It seems that theoretical cartographers and projectors were, if anything, encouraged by these failures. In 1755, a series of maps of Canada published in Paris showed a *Mer de L'Ouest* (western sea) extending itself as far as present-day Saskatchewan. It took another few decades to dispel the myth of its existence. In the aftermath of the fall of Quebec and the Seven Years' War (1756–1763) Britain was left unfettered to expand its empire across North America, but it first chose to attempt to navigate its way around it. Renowned British surveyor James Cook charted some of the Atlantic and Pacific coastlines. He too sought a passage around the top of the continent, but after sailing up the coast of North America through the Bering Straits to the Arctic Ocean he concluded that such a navigable passage and a *Mer de L'Ouest* did not exist.

By the end of the eighteenth century, civil surveys had progressed rapidly in eastern North America—George Washington had become a surveyor in 1748 at the tender age of sixteen. He opened the first survey into Virginian frontier—but for the most part even the near western interior was not yet cartographically secured. At the behest of the

BEFORE THE FALL OF QUEBEC AND THE END OF NEW FRANCE IN 1759, PIERRE GAULTIER DE VARENNES ET DE LA VÉRENDRYE AND HIS SONS EXPLORED THE INTERIOR AS FAR AS THE ROCKIES, SOMEWHERE SOUTH OF 49°N LAT.

Penns (Pennsylvania) and Calverts (Maryland), Englishmen Charles
Mason and Jeremiah Dixon performed their surveys inland from the
Atlantic Coast over a five-year period in the 1760s. The US American
Land Ordinance of 1785 was the first on the continent to adopt what
would be known in the British North American West a century later as
the "grid" survey; that is, rather than "metes-and-bounds surveys
(unconnected to any overall system of coordinates) that prevailed in
many of the seaboard states," the Land Ordinance "adopted the principle of rectangular survey" by which the land beginning at the Mason-
Dixon line and extending up the Ohio/Allegheny River beyond
Pittsburgh was to be "divided into townships 6 miles square, each
containing 36 sections of 640 acres each, 1 square mile."[5]

As for the Rocky Mountains, our ultimate interest in this book, their
existence was known but no maps had yet shown any details of them. In
1778, the HBC hired Philip Turnor (1751–1800) to survey the river routes
across Rupert's Land, and he in turn trained Peter Fidler (1769–1822) and
David Thompson (1770–1857) in the art of surveying. Thompson, the
continent's pre-eminent early western surveyor, created the first comprehensive map of the western interior including the Rocky Mountains. He
joined the HBC in 1784 and worked for several years as a trader on the
shores of Hudson Bay and on the Saskatchewan River. Taught astronomy
and the basics of surveying by Turnor in 1788, he laboured for another
decade under the HBC's lack of interest in surveying before leaving the
concern and joining its rival, the North West Company. The NWC
wanted Thompson to determine where the international boundary
between the American and British wests lay in relation to its fur posts,
but Thompson also surveyed extensively both north and south of the
boundary. During his decade and a half with the NWC he continued to
work as a trader while making his surveys. In the company of his Métis
wife, Charlotte Small, and some of their thirteen children, Thompson
travelled through the main passes of the Rocky Mountains gathering
detailed information with his compass, ten-inch brass sextant, and
artificial horizon. He was the first white man to use the upper Athabasca
River valley and Athabasca Pass, now part of Jasper National Park, to
cross over to the Pacific Slope and the Columbia River watershed. He did
so on 10 January 1811, just over a century before Bridgland made his
phototopographic survey of the Athabasca valley.

In 1812, Thompson retired to the Montreal area to compile fifteen
years' worth of sketches and survey notes. It took him nearly two years of

THOMPSON TRAVELLED
THROUGH THE MAIN
PASSES OF THE ROCKY
MOUNTAINS GATH-
ERING DETAILED
INFORMATION WITH HIS
COMPASS, TEN-INCH
BRASS SEXTANT, AND
ARTIFICIAL HORIZON.

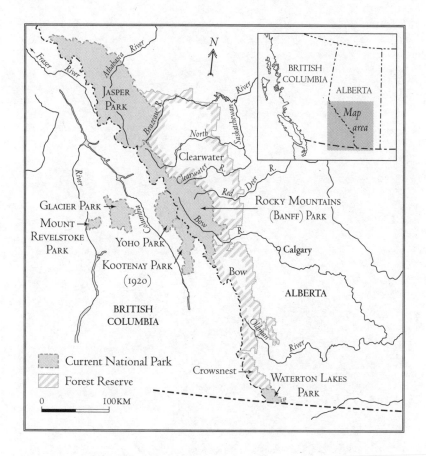

The Regions of Bridgland's Career as a Mountain Surveyor

calculations and drafting before, on 10 June 1814, he had completed his *Map of the North-West Territory of the Province of Canada From actual Survey during the years 1792 to 1812*, which he presented to William McGillivray of the NWC at its Montreal headquarters. Its size stupefied McGillivray. "Six feet, nine inches tall and ten feet, four inches wide and made of twenty-five sheets of paper glued together," it amounted, in Thompson's own retrospective estimation, to no more than "'a hasty, rough map.'"[6] Although there are a few gaps in the map, the accurate rendering of most major rivers and inland bodies of water from Lake Superior to the Pacific Ocean testifies to the thoroughness of his work. Few of his contemporaries would have agreed with his overly critical self-assessment.

The accuracy of Thompson's map is all the more remarkable given the relatively crude methods by which he estimated distances and the

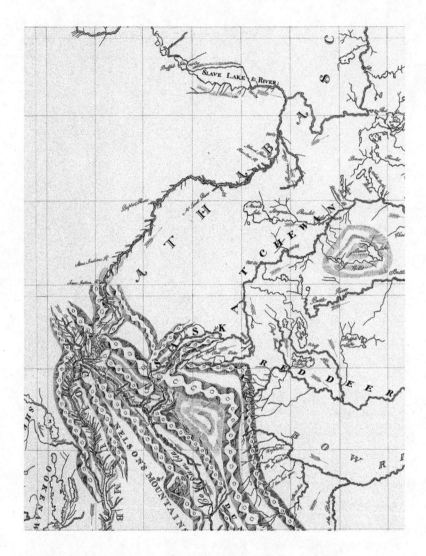

Facsimile of David Thompson's *Map of the North-West Territory of the Province of Canada From actual Survey during the years 1792 to 1812* (1814). This detail shows the emergence of the Bow, Red Deer, (North) Saskatchewan, and Athabasca rivers eastwards from the mountains, as well as the Big Bend of the Columbia River to the west of Athabasca Pass and what Thompson called the Nelson Mountains. Courtesy Fort William Historical Park, Thunder Bay, Ontario.

position of landmarks relative to one another. Unfortunately, this map was not disseminated to the public until after Thompson's death in 1857. Even as late as 1859, Sir George Simpson, the inland governor of the HBC, stated that Thompson's charts and notes were probably "worthless."[7] Aaron Arrowsmith, the famous Victorian publisher of maps, had by this time copied much of Thompson's work and published it in London as his own without crediting its source. Still, with Thompson the cartography and surveying of western Canada had enjoyed an auspicious beginning, with standards of accuracy set high for those who followed in his wake. And it was Thompson whose careful work put paid to the illusion, concocted after the mouth of the Columbia River had been discovered 11 May 1792 by Boston trader Robert Gray and explored to the vicinity of Mt Hood in October of that year by Lieutenant William Broughton of Captain George Vancouver's Royal Navy expedition, that because the lower Columbia River and the headwaters of the Missouri River lay in the same general latitude, only a short portage probably separated the two waterways. Of course, between 1792 and Thompson's completion of his map in 1814, Merriwether Lewis and William Clark's thirty-three-member expedition of 1804–1806 had succeeded in travelling to the Pacific Ocean by way of the upper Missouri and lower Columbia rivers, thereby, after Alexander Mackenzie in 1793, being the first whites to cross the continent. But Thompson was the first surveyor/cartographer to link the entire West. The information his map provided proved a tremendous boon to the NWC but, had it been more widely dissemi-nated sooner, its benefits would have been much greater and better known. As it was, when, in the same year (1814), the narrower Lewis and Clark map was published in Philadelphia, it eclipsed Thompson's much greater achievement and gave citizens of the United States a linear dream of a westering nation. Even though it took until 1859, two years after Thompson's death, before Gouverneur Kemble Warren (1830–1882) published the first complete map of the American West, the tremendous importance of maps for the very idea of 'West' that most North Americans held and still hold was beginning to be felt.[8]

After Thompson crossed Athabasca Pass in January 1811, the next survey of the Athabasca River valley occurred only fifteen years later. In 1826, Royal Navy lieutenant, hydrographer, and surveyor Æmilius Simpson (1792–1831) joined the HBC and made a survey of the normal transcontinental route from York Factory on Hudson Bay to Fort Vancouver, at tidewater on the lower Columbia River, by way of Fort Edmonton, the Assiniboine Portage from Fort Edmonton to the Athabasca River at Fort Assiniboine, and Athabasca Pass to the

THE TREMENDOUS IMPORTANCE OF MAPS FOR THE VERY IDEA OF 'WEST' THAT MOST NORTH AMERICANS HELD AND STILL HOLD WAS BEGINNING TO BE FELT.

Columbia River.[9] His work suffered from troubles with his chronometer and difficulties in determining magnetic variation, which, as he well realized, fell farther and farther out of line as his trip continued. On 7 October, he measured the elevation of Roche Miette (his Millette Rock) at 3,755 ft above the level of the Athabasca River, and guessed that its height above sea level was 13,000 ft (its actual height is only 7,600 ft [2,316 m]), but his measurements for longitude and latitude were fairly accurate. His early assessment of the arduous challenge of surveying in the mountains is telling:

> Altho' I very much regret the loss of the Mountain Barometer, yet circumstanced as I am it could not be applied to any general practical purpose, as it would be necessary to gain the Summits of the Mountains to ascertain the height by its means which would require considerable time and great labour, and indeed many if not most of them are inaccessible—[10]

But his doubts over the potential for surveying the region are balanced by his aesthetic appreciation in a delightful verbal survey while he is camped at the base of Roche Miette:

> The view now in all directions presents a continued mass of snow clad Hills towering their lofty summits in successive ranges their outlines assuming a great variety of forms, giving to the whole scene a grandeur and novelty beyond my power to describe—but is truly sublime....I certainly think our Encampment might vie in point of romantic appearance with many of far greater celebrity—On our left is the perpendicular face of this stupendous Rock rising to an elevation of upwards of three thousand feet its shadow casting a gloom over the deep defile, so opposite to the brilliant sky immediately ovre [sic] us that the mind feels an impression as if this situation was somewhat supernatural.[11]

Other than fur trade companies, the American army became an important agent of the surveying and mapping of the continent beginning in 1838, with the establishment of the Corps of Topographical Engineers.[12] In British North America, the agency that made significant early contributions was the Geological Survey of Canada (GSC). Formed in 1842 under the directorship of Sir William Edmond Logan

(1798–1875), the first Canadian-born person to be knighted by a British monarch, the GSC took as its mandate the compiling of an inventory of geoscientific information, which might encourage natural resource development in the British North American colonies. This was a time of rapid development. Expansionist-minded businessmen and entrepreneurs in Canada West (Ontario) were turning their sights towards Rupert's Land as a possible source of wealth. Coal to fuel the nascent system of railways was most prized on a continent where the conquering of space by the technology of the steam locomotive seemed essential to a prosperous future. When Logan informed the public that no coal existed within the Province of Canada, expansion of that jurisdiction to whatever unclaimed regions of the continent *had* coal soon dominated the minds of capitalists and politicians alike. The potential of the territory under the control of the HBC had to be assessed.

Canadians did not stand alone in their curiosity over the potential of the northwestern interior. In 1857 two separate expeditions, one British and one Canadian, were mounted to explore the western interior. Captain John Palliser (1817–1887) led the British expedition. He spent three summers travelling through the prairies and Rocky Mountains to the Pacific coast. With the help of the famous botanist Sir James Hector (1834–1907) and others, Palliser collected astronomical readings, meteorological information, and geological and botanical data. Like the Canadian one, the British expedition was mounted as a scientific mission, intended to provide an inventory of the natural resources, climate, flora, fauna, and soil quality and other geography of the continental hinterlands. Such an inventory would determine whether or not it was economically worthwhile to retain the rights to this vast and only sparsely populated land, and whether or not it would serve the British government to retain governance of the area.[13]

Hector visited the Athabasca valley at Roche Miette on the last day of January 1859. For three decades, the location of the fur trade post known as Jasper House (also known as Jasper House II) had existed across the Athabasca and upstream from Roche Miette (22.5 km south of the location of Jasper House I, or Rocky Mountain Fort, which had been on the west shore of Brûlé Lake from 1813 to 1829). Hector located it at 53°12'15"N Lat. (its location is currently given as 53°09'N Lat., 117°59'W Long.), and "found that the Roche Miette, which seemed almost to overhang the fort, is nearly at a distance of four and

EXPANSIONIST-MINDED
BUSINESSMEN AND
ENTREPRENEURS
IN CANADA WEST
(ONTARIO) WERE
TURNING THEIR SIGHTS
TOWARDS RUPERT'S
LAND AS A POSSIBLE
SOURCE OF WEALTH.

a half miles, while its summit is elevated 5,800 feet" above the floor of the valley, which he found was 3,762 ft. above sea level.[14] With his mixed-blood Iroquois guide, Tekarra, Hector continued up the valley south and west, although he did not divert from the principal fur trade route or, for that matter, pursue it as far as the height of land at Athabasca Pass (his guide's ailing foot and bad weather were to blame for his decision to turn back after seeing only the confluence of the Whirlpool and Athabasca rivers). However, he saw much of the country that Bridgland would survey fifty-six years later and noted its mineral and timber potential, in particular.

Unlike Palliser's expedition, the Canadian expedition was not mandated to extend as far as the mountains. Simon James Dawson (1820–1902) and Henry Youle Hind (1823–1908) led it in 1857 and 1858 in the vast region between Lake Superior and the prairies as far as the middle of modern-day Saskatchewan. This survey gathered topographic information as well as a wealth of botanical and geological data. The men also drew up plans for a road-and-water link between the western end of Lake Superior and Fort Garry (now Winnipeg), which was completed about a decade later and was called Dawson's Road. At the time of their surveys, the vision of a railway extending from St Paul, Minnesota, into Manitoba promised greatly to increase the flow of immigrants to the area. (That vision was not realized until December 1878, the month and year of M.P. Bridgland's birth.[15]) Owing to the rough terrain of the Canadian Shield north of Lake Superior, the only easy route to Manitoba lay through the United States. This fact concerned some expansionists, in particular, those who were not yet at ease with American neighbours who had only grudgingly ceded their claim to the entire Pacific Slope in the Oregon Treaty of 1846, and who, on the brink of civil war, appeared precarious fellow residents on the continent. Canadians began to consider seriously the need for roads or rail lines that would replace the long, lake paddles and punishing portages that then constituted the only all-British routes between the East and the West. By the 1850s, a parallel perception had come into focus, the idea of the West not as a wilderness but as a potential hinterland of resources, including real estate for an agrarian population in Ontario that had brought all arable land under cultivation and was looking for more. The idea of land as property began to fire the imagination of some in the East.

BY THE 1850S, A PARALLEL PERCEPTION HAD COME INTO FOCUS, THE IDEA OF THE WEST NOT AS A WILDERNESS BUT AS A POTENTIAL HINTERLAND OF RESOURCES, INCLUDING REAL ESTATE FOR AN AGRARIAN POPULATION IN ONTARIO THAT HAD BROUGHT ALL ARABLE LAND UNDER CULTIVATION AND WAS LOOKING FOR MORE.

❋ *The Advent of the Dominion Lands Survey* ❋

AFTER CONFEDERATION IN 1867, provinces retained control of their respective lands and survey structures, and federal survey activity was limited to the projects undertaken by the GSC. Not a province, the North-West Territories became a federal jurisdiction when the Rupert's Land Act, introduced in Parliament by William McDougall, the minister of public works in the first Dominion government, was passed on 31 July 1868. Actual transfer had yet to occur, however, and would not do so before violent confrontation occurred in which the act of surveying figured prominently.

In 1869, a federal program was initiated to survey the international boundary and to commence a skeleton survey of the territory north of it. From this survey, squatters' rights could be determined and treaties could be negotiated with Native peoples. In that year, McDougall assigned Lieutenant-Colonel John Stoughton Dennis (1820–1885) the task of "selecting the most suitable localities for the survey of Townships for immediate settlement" in the vicinity of the Red River Settlement and upper and lower forts Garry.[16]

Upon his arrival at Red River in August 1869, the controversial nature of his work dawned on Dennis: Métis farms were already laid out in the French method of survey, by which the land was divided into long strips, or *rangs*, with the narrow ends starting from the frontage on Red River. As it did in New France, this distribution allowed each landowner direct access to water and proximity to neighbours. The survey that Dennis had planned to implement followed the square-township used in the United States. While the square system was especially efficient for mapping large tracts of land consistently, its implementation could only disrupt a different existing system. Métis wanted assurances that the land granted them by the HBC would not be revoked when the North-West Territories was officially re-surveyed and transferred to the control of the Canadian government. Dennis, whose commission was aimed at surveying land for future settlers, aimed to work with residents already in place, but not all his men saw matters his way. On 11 October, one survey party had to stop work when armed Métis confronted it.

On 30 November, McDougall was in Pembina, on the American side of the international border. A "noted Canadian annexationist, and the man nominated to be first lieutenant-governor of the North-West

WHILE THE SQUARE SYSTEM WAS ESPECIALLY EFFICIENT FOR MAPPING LARGE TRACTS OF LAND CONSISTENTLY, ITS IMPLEMENTATION COULD ONLY DISRUPT A DIFFERENT EXISTING SYSTEM.

Territories," he had grown impatient at the slow response to the crisis by the government of Sir John A. Macdonald, but also was unaware that the transfer had been delayed by the prime minister because he thought it improper for Canada to be able to take over land that was not at peace with the Dominion. The unrest was, in his view, a problem for Britain to resolve before the transfer proceeded. However, McDougall, not knowing as much, prematurely and illegally proclaimed that the territories lay under Canadian jurisdiction. The citizens of the Red River Settlement heard about his declaration the next day.[17] Dennis abandoned his survey and left the settlement but some of his crew did not. One member of it, the infamous Thomas Scott, a twenty-eight-year-old survey labourer from Ontario, got himself arrested more than once. To Scott's astonishment, Louis Riel's provisional government made an example of him and had him executed on 4 March 1870. This marked the height of the first Riel Rebellion. The establishment of the Province of Manitoba in May averted further problems, but the survey did not resume until after the rebellion had died out and the transfer could occur peacefully.

THE ACQUISITION
OF THIS VAST
TRACT INCREASED
THE CANADIAN
GOVERNMENT'S
LAND HOLDINGS
EXPONENTIALLY,
FOR IT STRETCHED
FROM UNGAVA IN
HUDSON STRAIT TO
THE ROCKIES.

The HBC relinquished title on 15 July 1870 in return for £300,000 and land. The acquisition of this vast tract increased the Canadian government's land holdings exponentially, for it stretched from Ungava in Hudson Strait to the Rockies. The equivalent of about one-twentieth of the land formerly held by the company was retained through an agreement that, when the territory was all surveyed into the thirty-six square-mile townships of the grid survey being planned, the company would and did receive "section 8 and all of section 26 but the northeast quarter in every township. In every fifth township, however, the Company received all of 26."[18]

The Dominion Lands Survey (DLS) finally got under way again on 10 July 1871. Dennis was appointed surveyor general of the Dominion Lands Office (the Department of the Interior came into being two years later, and, with it, the Topographical Surveys Branch). Macdonald assigned him the completion of the survey, for the Canadian government planned to capitalize on its newly acquired lands and natural resources as it promoted both western settlement and the building of a railway; the survey was needed to achieve both aims. As well, of course, Macdonald had in mind his promise to British Columbia. That western colony joined Confederation on 20 July 1871 on the strength of the Canadian government's earnest to provide a transcontinental railway link within ten years.

The infrastructure necessary for the survey of the West called for a system that could identify and locate individual parcels of land. In 1872, the parliamentary approval of the *Dominion Lands Act* provided the legal framework for the distribution of land to settlers. It was modelled on American homestead legislation and stipulated the terms of the sale of land: a fee of ten dollars and the performance of specific settlement "duties" which included building a residence and farming a certain portion of the land annually. The government also sought information relating to exploitable natural resources and hired GSC surveyors to work in tandem with the DLS. But the surveyors were not yet a posse of salaried men; all of the surveying accomplished in the West during the 1870s and early 1880s was undertaken by men who had won tenders. It is a wonder that Dennis and his successor, Lindsay Russell, could meet the challenge at the heart of which was the necessity for consistent work.

IN 1872, THE PARLIAMENTARY APPROVAL OF THE DOMINION LANDS ACT PROVIDED THE LEGAL FRAMEWORK FOR THE DISTRIBUTION OF LAND TO SETTLERS.

Dennis had approved a system of survey that could be used for all Dominion Lands. The first *Manual of Survey* (1884) set out such a system: prairie lands were to be divided into townships six miles square. Townships comprised thirty-six sections. Each of these 640-acre sections was divided into 160-acre quarter sections. One quarter section became the standard size for a homestead. Around the perimeter of each township block, the initial system (later revised) also stipulated that a strip of land should be reserved as a road allowance. This feature reflected an alteration from the American grid and it exerted quite a difference on settlers:

> The existence of this shared highway [between townships] tended to draw houses toward it, shortening the distance between families. South of the forty-ninth parallel, the U.S. prairie farmer endured a uniquely solitary existence where progress depended on an individual's ability to impose his or her will upon the land.[19]

The *Manual of Surveys* included practical directions: for the establishment of governing meridians and base lines, instructions were included on proper azimuth use.[20] As well, for the delineation of blocks, townships, sections, and quarters, instructions were supplied for the proper use of a traverse with chains and the theodolite/transit.[21] Many of the men hired to carry out the survey had experience from previous but smaller and less systematic surveys in Ontario, Quebec, or the Maritimes, and others were graduating astronomy and mathematics students who saw great prospects for work in the West. The experienced hands, who surveyed by terrestrial magnetism, proved to be less well suited to the task than the grads in astronomy and maths because Dennis's method depended on the taking of astronomical readings, not compass bearings. To avoid inaccuracies in the control points of the principal meridians, Dennis formed around himself a Special Survey, which included, among others, Daniel-Édouard-Gaston Deville (1849–1924), William Frederick King (1854–1916), Lindsay Russell (1839–1912), Alexander Russell (1842–1922) and Otto Julius Klotz (1852–1923). Klotz served as the first president of the Association of Dominion Land Surveyors when it formed in April 1882. Most of the survey team went on to positions of prominence in the Department of the Interior, and enjoyed international reputations for the standard of their work and their contributions to the development of surveying.

This department was instituted in 1873 to control all matters of land survey in the North-West. By bringing together such agencies as the Dominion Lands Branch and the GSC, this new department effectively joined the forces of all types of surveyors under one governing body. In the first half of the 1870s, block and township surveys were conducted in Manitoba, gradually extending north and west from Winnipeg. At the same time, the Imperial International Boundary Commission extended its survey of the 49th parallel from Lake Superior to the Rocky Mountains.

From the outset, every surveyor employed an instrument called a theodolite by the British and a transit by American and Canadian surveyors. This permitted the estimation of straight lines on uneven ground. A chain, sixty-six feet of steel links each a little less than eight inches in length, was their measuring device. (The fact that it would both sag on uneven terrain and lengthen or shorten with variations in temperature guaranteed inaccuracies that present-day standards of mensuration would consider unacceptable.) Astronomical observations could verify the men's work on the ground. A survey crew consisted of a chief, his assistant, a number of chainmen and axemen, and a cook. It was early November 1879 before any group of the Dominion Lands Survey had worked as far west as the area of Fort Edmonton.[22] (Earlier in that year competing surveys in the United States were brought by Congress under one civilian body and given the name of the United

States Geological Survey.) In 1880, Montague Aldous was able, though just barely, to complete the staking of the fifth meridian along 114°W Long. The work was comparatively simple on the land west of Edmonton, but, as they worked their way south during the summer, he and his party hit the Porcupine Hills, a swollen Belly River, and the mountains west of Pincher Creek and present-day Waterton Lakes National Park.[23]

Still, gradually, one of the largest land masses ever surveyed in the world was being brought under identification by the square-mile grid survey. Comparably-sized parcels of land bearing a clear geographical relationship one to another were coming into being across the West: "all rural addresses in western Canada were established according to a single principle" because the numbering of townships and range roads depended from the international border (49°N latitude) and the first meridian (97°27'28.4"W Long.) and, as the survey spread west, its six successors: 102°W, 106°W, 110°W, 114°W, 118°W, and 122°W Long.

The DLS took the necessary first step in the settlement of the Canadian West. That step was in reality several steps: a first survey (1871–79) and four subsequent surveys, including by far the largest, the third, which began in 1881 and continues to the present, in which townships measure 486 chains x 483 chains (about 36.67 mi^2 or 95 km^2).[24] Correction lines were included in the surveys, of course, to account for the curvature of the Earth (no more level on the prairies than elsewhere, popular perceptions notwithstanding). Thankfully, Jared Mansfield (1759–1830), appointed surveyor-general by American president Thomas Jefferson in 1803, had initiated the "jog" in his surveys of Indiana and Illinois in the first decade of the century:

> One of the long-recognized problems...was the conflict between rectangularity and convergency. North-south lines in surveying are not in the long run parallel. They cannot be because they converge as they run northward toward the pole, as the meridians do on a globe. Therefore, without adjustments, it would be impossible for a six-mile-square township to be perfectly square or to be uniform in size compared to an adjacent township to the north. ...[Mansfield] devised a compromise between rectangularity and convergence that became the practice. By his concept the principal meridians and parallel base line, the broad framework of the survey, would remain uninterrupted lines.

STILL, GRADUALLY, ONE OF THE LARGEST LAND MASSES EVER SURVEYED IN THE WORLD WAS BEING BROUGHT UNDER IDENTIFICATION BY THE SQUARE-MILE GRID SURVEY.

But, within that framework, all the subordinate meridians would take a slight jog at each subordinate parallel and then start northward again the same distance apart as they were at the previous parallel. Such a map of township boundaries took on the appearance of off-line masonry.[25]

In 1884, a significant milestone in surveying techniques was reached with the international implementation of geographical time zones, agreed upon at the Prime Meridian Conference in Washington. This innovation, first proposed by Scottish-Canadian Sandford (later Sir Sandford) Fleming (1827–1915) as a solution to the problems in regulating train schedules, also made it easier for surveyors to locate meridians with greater accuracy. So did the telegraph; by it, accurate time signals could be transmitted long distances more or less instantaneously (in the case of the grid survey they were transmitted first from the observatory in Chicago). Surveyors would still have to determine mean solar time, but no longer had need of chronometers because they could now calculate the difference between the standard time and local mean solar time (and thus longitude) based on the declination and known apparent rate of the sun's movement at any given latitude.[26]

❋ *Railway Survey in the Athabasca Valley, 1872* ❋

WELL BEFORE 1884, concerns over the railway had arisen. The first transcontinental railroad (the American term) had opened across the United States on 10 May 1869, when the golden spike was driven at Promontory, Utah; the American surveys for rail and settlement had preceded either the need or the effort to survey for a transcontinental railway north of the international border. Despite a recession soon after Confederation and a consequent lack of investment in railways, the plan remained that a line would be built across Canada as soon as possible. Captain Palliser's decade-old report from 1863 to the British government that the southern prairies were arid and unsuitable for farming led the Canadian government to appoint Fleming to supervise surveys across the prairies and through the Rocky Mountains to determine the route to the Pacific Ocean most suitable for settlement and a railway. Fleming unequivocally recommended a route over Yellowhead Pass (up the Athabasca and Miette rivers to the Fraser River), noting its comparatively lush and broad valleys, gentle changes in elevation, and

DESPITE A RECESSION SOON AFTER CONFEDERATION AND A CONSEQUENT LACK OF INVESTMENT IN RAILWAYS, THE PLAN REMAINED THAT A LINE WOULD BE BUILT ACROSS CANADA AS SOON AS POSSIBLE.

generally low elevation.[27] It is precisely the combination of low gradient and aesthetic attractions that the popular narrative about the work of Fleming's survey party noted in describing the Athabasca valley. "We were," wrote the expedition's secretary on 11 September 1872, "entering the magnificent Jasper portals of the Rocky Mountains by a quiet path winding between groves of trees and rich lawns like an English gentleman's park. … We could now sympathise with the daft enthusiast, who returned home after years of absence, and when asked what he had as an equivalent for so much lost time,—answered only 'I have seen the Rocky Mountains.'"[28] George Monro Grant (1835–1902), the expedition's secretary and the future principal of Queen's University, waxed poetic about the virtues of Roche Miette:

A GOOD PHOTOGRAPHER
WOULD CERTAINLY
MAKE A NAME AND
PERHAPS A FORTUNE,
IF HE CAME UP HERE
AND TOOK VIEWS.

> Myette is the characteristic mountain of the Jasper valley. There are others as high, but its grand bare forehead is recognized everywhere. It is five thousand eight hundred feet above the valley, or over nine thousand feet above the sea.
>
> A good photographer would certainly make a name and perhaps a fortune, if he came up here and took views. At every step we longed for a camera. …the most wonderful object was Roche à Myette, right above us on our left. That imposing sphinx-like head with the swelling Elizabethan ruff of sandstone and shales all around the neck, save on one side where a corrugated mass of party coloured strata twisted like a coil of serpents from far down nearly half way up the head, haunted us for days. Mighty must have been the forces that upreared and shaped such a monument. Vertical strata were piled on horizontal, and horizontal again on the vertical, as if nature had determined to build a tower that would reach to the skies. As we passed this old warder of the valley, the sun was setting behind Roche Suette [*sic*: Roche DeSmet]. A warm south-west wind as it came in contact with the snowy summit formed heavy clouds, that threw long black shadows, and threatened rain; but the wind carried them past to empty their buckets on the woods and prairies.[29]

Once again, an earlier surveyor than Morrison Parsons Bridgland remarked particularly on Roche Miette and the Athabasca valley, again exaggerating the mountain's height, this time fancifully personifying it

Photograph by Charles George Horetzky of the view of Jasper House II and Roche Miette, Jan. 1872.
Courtesy Glenbow Archives, NA–1408–12.

in courtly Elizabethan dress, and calling for the very instrument—the camera—to depict it that Bridgland would deploy in his survey just over forty years later.[30]

But this was a representation of the mountains that a book aimed at a popular readership could be expected to provide. For surveyors in search of a viable railway route, what did it matter that a valley held aesthetic charms? No engineer himself but rather a publicist, at least in his capacity as the expedition's secretary, Grant had a very different perspective from the one that a surveyor necessarily adopted. To gain that other, nose-to-the-ground perspective, one needs to read the diary of a railway surveyor. The one kept by Walter Moberly (1832–1915), an engineer from British Columbia who worked in association with but independently from Fleming on his survey, tells a fascinating story about his leadership of the survey of the Athabasca River valley from November 1872 to February 1873.

Engraving of Charles George Horetzky's photograph in George Monro Grant, *Ocean to Ocean* (1873), facing 231.

As the chief engineer in charge of Party S, Moberly began his railway survey from the West Coast on the day on which British Columbia entered Confederation, 20 July 1871.[31] In 1872, his second year of surveying, he was meant to rendezvous with Fleming's westbound survey around Jasper House and concentrate on the Yellowhead Pass (Fleming's choice over Moberly's favoured route, Howse Pass). In the event, the two met up at the confluence of the Miette and Athabasca rivers (just upstream from the town site of present-day Jasper) on 14 September. Thereafter, Moberly and his crew spent much time ignoring Fleming's subsequent order to discontinue surveying for the season and, instead, conducted a transit and levelling survey in search of the most practicable route from the height of land in Yellowhead Pass in the west to the foot of the easternmost mountain of the Athabasca valley in the east.[32] To read his unpublished account of the toil involved in the survey disabuses one of any romantic illusions of the beauty of mountainous terrain to a surveyor. Moberly had great difficulty deciding which bank of the stretch of the Athabasca that today lies inside Jasper National Park and downstream of the town site of Jasper was better suited to a railway's needs. At Brûlé Lake (the widening of the Athabasca River just outside the park's eastern boundary), he found himself exasperated by its eastern shore's "high sand ridges with deep rifts between that have been cut out by the strong and almost incessant winds that blow in this most disagreeable Valley." The sand dunes, an unprotected area in the process today of being permanently altered by recreational drivers on their ATVs (all terrain vehicles), "presented a most unfavourable locality for a railway, a line built here would have to be covered the entire distance."[33] Moreover, he was staggered to find that, even in such a broad valley, two bridges across the river and likely a tunnel were going to be necessary.

ONE OF THE
OBSTACLES WAS OLD
ROCHE MIETTE, ITS
DISTINCTIVE PEAK A
LOVELY SIGHT TO
BEHOLD FROM A
DISTANCE PERHAPS
BUT ITS FOOT A
STUBBORN IMPEDIMENT
TO TRAVELLERS.

One of the obstacles was old Roche Miette, its distinctive peak a lovely sight to behold from a distance perhaps but its foot a stubborn impediment to travellers; even today's passing motorists, benefiting from the work long ago of a good bit of dynamite, can see how its lower portion, known as Syncline Ridge, angles sharply down into a fast-flowing outer bend of the Athabasca, leaving no level ground for a railway line. On 18 December, Ashdown Green, the transit man on Survey S, recorded simply that he ran his survey "to a bluff of perpendicular rock over which we could not get." Thereafter, the foot of Roche Miette simply gets called "the rock" or "the bluff," so distinctive an impediment was it to a surveyor.[34] Earlier, on the last day of November, after taking up residence in the area and being hospitably received by the manager of Jasper House (mixed-blood Iroquois Paulette Finley, his family, and relatives), Moberly got himself figuratively tongue-tied trying to sort out the options:

> My object in crossing the Athabasca river and taking the line on this side is that it will be a shorter line by 2 to 2½ miles between Athabasca Depot [about opposite and .5 km upstream from the mouth of Maligne River, that is, on the same side of the river as, and several kilometres downstream from, the town of Jasper] & Fiddle R. [the first river that motorists cross after entering the park from the east on the Yellowhead Highway] if taken on the left bank of the Athabasca river & crossing by bridge some 2 or 2½ miles below Jasper House. The bridge at the latter point would also have to be much longer than where above proposed, there are several high projecting points & much heavy work where a great length of tunnelling wd. be required along Jasper Lake if

line should be kept on left bank [the highway runs along the right bank]. In addition to the three rocky points on right bank of Athabasca river described there are also other objections to the line I have followed viz. the crossing of the Rocky River opposite Jasper House & the long bluff below it which projects from the Roche de Myette. Whichever bank the line s[houl]d. ultimately follow some heavy work will have to be encountered. I think however mile for mile the difference of the cost of the road would not be material, the difference in the distance being in favour of the line I have surveyed. I however am certain that a survey should also be made on the opposite bank before finally deciding wh[ich]. shall be adopted. As I was convinced that this would be requisite I decided to make the survey at present along the shorter of the two.[35]

Part of the irony in this passage today lies in the fact that, in the short term, all Moberly's work went for nought in that neither bank was chosen because Fleming's recommendation of the Yellowhead Pass was not adopted. But an even richer irony is that, when rail service opened the Athabasca valley forty years after Moberly agonized over the matter, lines were constructed up *both* banks. By the time of Bridgland's survey in 1915, both the Canadian Northern and Grand Trunk Pacific railways were built at that point in the valley. The Gordian knot that Moberly first articulated was cut in a way that he probably could never have imagined.[36]

By 1880, the skeleton survey grid being worked by the Dominion Lands Survey had progressed as far north and west as Edmonton along the planned route for the railway line. Within a year, however, the Canadian Pacific Railway (CPR) overtook and changed the direction of the survey. Up until this time, it was aimed at covering the area through which it was thought the transcontinental rail line would run; after all, first settlements were likeliest in the regions served by rail. But the *Canadian Pacific Railway Act*, which received royal assent 15 February 1881, had turned the project over to the chartered company of that name, over which five Montreal Scots—George Stephen, James Jerome Hill, Duncan McIntyre, and Richard Bladworth Angus—presided, with Donald Alexander Smith (Lord Strathcona and Mount Royal, the man who would name Banff in 1883 after his Scottish home town) and Norman Kittson involved as investors. They had a different plan in

BY 1880, THE SKELETON SURVEY GRID BEING WORKED BY THE DOMINION LANDS SURVEY HAD PROGRESSED AS FAR NORTH AND WEST AS EDMONTON ALONG THE PLANNED ROUTE FOR THE RAILWAY LINE.

mind because they considered Fleming's recommendation of the route through the upper Athabasca River valley and Yellowhead Pass too expensive. Instead, they investigated and then chose a shorter if steeper southerly route through the mountains. This decision was influenced in part by the report written by John Macoun (1831–1920), a Dominion botanist with the GSC, that the southern prairies, known as the arid Palliser's Triangle, were in fact full of lush grazing lands. The bulk of the subdivision surveys conducted along the railway traverses were diverted south where a decade later the CPR would complete the building of the transcontinental line through what would become Regina and Calgary, rather than Saskatoon, Edmonton, the Athabasca River valley, and Yellowhead Pass. Roche Miette would remain unsurveyed until subsequent railways were constructed in the twentieth century.

✳ *Surveyors Strive to Stay Ahead of the Railway Builders* ✳

LINDSAY RUSSELL, who served as surveyor general from 1878 to 1884 and as deputy minister of the Department of the Interior for eighteen months from 1882 to June 1883, and Édouard-Gaston Deville, chief inspector of the Surveys Branch, scrambled to prepare for the influx of settlers they anticipated in conjunction with the advent of the new CPR route. This grid survey, everyone was realizing, required an immense amount of manpower with hundreds of surveyors trying to keep ahead of the thousands of men who moved the end of steel steadily westward: "sometimes the ink was scarcely dry on their plans when the army of tracklayers was at their heels."[37] In the event, a lack of experienced surveyors for the amount of work available meant that in 1883, when parts of the railway reached Alberta ahead of the township surveys, newly arrived immigrants had no choice but to settle on unsurveyed land without any assurance that they would later be able to secure the legal title to it. The fact that the surveys were not progressing as rapidly as railway construction concerned the CPR, which advised Prime Minister Macdonald that it would hire surveyors of its own to complete the subdivisions required to determine the boundaries of its land grant. The Surveys Branch opposed this development and prevailed. Deville was largely responsible for persuading Macdonald that only DLS surveyors who had been properly trained and certified should complete the task. The errors of calculation that he discovered in many privately surveyed township boundaries convinced both men that independent surveyors worked less carefully than their peers in the DLS.

The North-West Rebellion temporarily interrupted the progress of the prairie surveys in March 1885. For a second time, surveyors put down their instruments and took up arms against a Métis uprising. Two months later, Louis Riel's forces surrendered. Three defining moments in Canadian national history then occurred in the penultimate month of the year: the last spike in the CPR was driven at Craigellachie, British Columbia, on Saturday 7 November; nine days later, on Monday 16 November, Riel was hanged in Regina; another nine days later, on Wednesday 25 November, a federal order-in-council established Banff Hot Springs Reserve on ten acres (four hectares), the nation's first park outside a city limits. Thereafter, settlers, tourists— already by 1888 more than five thousand had thronged to the Cave and Basin to take its waters[38]—government agents, investors, and developers came to the mountains in increasing numbers, stirring interest in the land adjacent to the railway, and spurring the Department of the Interior to produce an inventory of the natural resources, flora, fauna, and geographic features of the mountains. This work had to occur, however, at the same time as other needs made their presence felt on the work of the DLS.

THREE DEFINING MOMENTS IN CANADIAN NATIONAL HISTORY THEN OCCURRED IN THE PENULTIMATE MONTH OF THE YEAR.

In fact, surveyors were suddenly urgently required at the end of the century. As a result of the Klondike Gold Rush (1896–99) and Alaska Boundary Dispute (1898–1903), both the Alaska/British Columbia boundary and the Alaska/Yukon boundary had to be surveyed. Since the St Elias Range runs through the former and it, together with the Ogilvie Range and the Brooks Range-British Mountains, run through the latter, phototopography was much in use in the North under the superintendence of James Joseph McArthur, DLS.[39] The last decade of the century also witnessed much activity in the Rocky Mountains, as well as on the prairies, wherever a demand was evident either for railways, settlement, irrigation assessment, or the development of resources or infrastructure. Several prominent DLS surveyors, including Arthur Oliver Wheeler (1860–1945) and William Pearce (1848–1930), had been studying American methods of routing water to arid areas. They soon concluded that in order to irrigate southern Alberta and Saskatchewan a system of reservoirs and canals would have to be established. The *North-West Irrigation Act* was passed in July 1894 to reserve lands for such infrastructure, and to ensure the water rights of the Dominion government. In anticipation of its approval by Parliament, the Topographical Surveys Branch had launched the Canadian Irrigation Survey in June. Detailed records of watercourses, their sources, and flow rates comprised this survey's aims. Perhaps most significantly, it introduced the production of contour maps into the work of the DLS in the West, for maps showing contours were vital for the determination of which waters could be diverted where by the use of gravity alone, and where reservoirs could be established. Two divisions of the survey went about their work. In 1895, when M.P. Bridgland was in the midst of his high school years, Wheeler's division entered the foothills south and west of Calgary to survey ground extending from the watersheds of the Bow River in the north to the Waterton River in the south. In this rougher terrain, where vertical variation presented a constant challenge to the art of surveying, surveyors working west of the prairies began to use cameras. They had, however, begun to be used in the mountains several years earlier by another, legendary DLS surveyor.

❋ *Surveying the Canadian Rocky Mountains Using Phototopography* ❋

HISTORICAL ACCOUNTS OF SURVEYING IN CANADA make a good deal of the 178 million acres of prairie land measured and plotted in the first fifty years after Confederation. As the largest ever completed under a single integrated system, the DLS survey is indeed "one of the great civil engineering triumphs of all time."[40] The sheer magnitude of the project is evident in astronauts' observations that the Great Wall of China and the regular land division pattern over much of North America mark the only artificial features on Earth visible from outer space. That over one million quarter-section-sized homesteads could be mapped in such short time testifies to the efficiency of a uniform, checkerboard-style survey and to the hard work of thousands of dedicated surveyors. But this feat also owes its success to the simple geographical fact that, for the most part, the Prairie provinces offered few topographical barriers to the overall westward progress of crews and their survey equipment.

Two reasons accounted for the difficulty that the westward push of township surveys met at the foothills of the Rocky Mountains: the mountains could not be efficiently surveyed with traditional methods, and the Department of the Interior realized that the mountainous areas

AS THE LARGEST EVER COMPLETED UNDER A SINGLE INTEGRATED SYSTEM, THE DLS SURVEY IS INDEED "ONE OF THE GREAT CIVIL ENGINEERING TRIUMPHS OF ALL TIME."

of the country were more valuable for their potential natural resources and tourism than for homesteads. As to the former, the system of measurement by chain was relatively efficient in flat areas, but hilly and tree-covered areas and bodies of water required that the surveyors apply corrections to their horizontal measurements when they encountered any considerable changes of elevation. These corrections could not amount to more than educated guesses. As to the latter, moving from prairie to mountain, surveyors, like Wheeler on the Canadian Irrigation Survey, had to change their work from staking the boundaries of homesteads to making inventories of resources and detailed maps of the land's traversable routes. Both reasons meant that, somehow, the third dimension, altitude, had to be represented in a system of symbols comprehensible not just to a surveyor but also to a tourist, entrepreneur, investor, or other occasional user. The grid system of surveying could not accommodate this requirement.

In 1759, nearly a century before the advent of photography, German mathematician Johann Heinrich Lambert proposed that one could create accurate topographic charts from a series of views of the landscape by using the inverse principles of perspective. The system was tested some thirty years later, when Charles François Beautemps-Beaupré produced charts of unusual quality from a series of freehand sketches of South Australia made during an expedition led by French navigator Joseph Antoine Bruni d'Entrecasteaux. In the nineteenth century, photographic methods improved from the eight-hour exposure time that Joseph Nicéphore Niépce took in 1826 to make the first photograph, and the tens of minutes to which Louis Jacques Mandé Daguerre reduced the process in 1837. As the methods improved and their accuracy in showing the spatial relations between objects in a landscape grew better appreciated (especially by French geodesist Dominique François Jean Arago), so did the practicality of iconometry. Coined only in 1898, iconometry referred to what in essence is the opposite of perspective drawing; that is, it refers to measuring (rather than free-hand drawing) dimensions of objects from their perspectives. Soon it came to refer as well to those graphic constructions that serve to convert perspectives into horizontal projections. Fortunately for Canadian surveyors, the advancement of photography and iconometry more or less coincided with the arrival of DLS surveyors in the western mountains.[41]

By 1850, French colonel Aimé Laussedat, the grandfather of modern photographic surveying techniques, had formulated a way by

FORTUNATELY FOR CANADIAN SURVEYORS, THE ADVANCEMENT OF PHOTOGRAPHY AND ICONOMETRY MORE OR LESS COINCIDED WITH THE ARRIVAL OF DLS SURVEYORS IN THE WESTERN MOUNTAINS.

which to calculate measurements of perspective from sketches he made with the aid of a camera lucida, a projector-like precursor to the modern camera. His first experiments with photography and surveying occurred two years later, when photographs were still made through a painstaking "wet process" in which the photographer coated a small rectangular plate of glass with an emulsion, then exposed and developed the plate before the emulsion dried and lost sensitivity. It was this laborious technique that Humphrey Lloyd Hime (1833–1903) had striven to deploy during the Canadian government-sponsored survey of the near prairies under Henry Youle Hind in 1858, but Hime was chiefly only a press photographer, not a topographical surveyor.[42]

By 1873, the commercial availability of dry plates made photography more mobile and fieldwork simpler, although many surveyors and photographers had already begun to conduct field experiments with the wet plates. The Corps of Royal Engineers had a photographic studio installed at their headquarters in Chatham, England, by June 1856, and training in photographic methods was given to sappers from 1857 onwards. So when the Corps came to North America to mark the international boundary between British Columbia and the United States, photographers, if not yet phototopographers, numbered among the Royal Engineers. The use of photography was not considered a success in this project, however, and survey work waited more than a decade before photography made its presence felt in a more regular respect, but when it did, it can be fairly said, it marked the only great advance on plane table surveying (used by the Romans) other than the invention of an accurate theodolite/transit. On the boundary survey along the 49th parallel between the American plains and the Canadian prairies from 1872 to 1874, photographers illustrated the surveyors' work, although clouds of grasshoppers precluded much picture-taking for several weeks. The 250 photographs served chiefly to illustrate the boundary commissioner's report, that is, in a documentary rather than a phototopographical capacity.[43] That latter usage would come prominently into service in the 1880s as the grid survey proceeded west and the topography began offering up what the camera could best record, that is, "boldly accentuated areas, such as are found in all mountains regions."[44]

Perhaps Laussedat's most important contribution to the advancement of phototopography was his method of surveying, which combined a ring of photographs with triangulation from a known

...IT [PHOTOGRAPHY] MARKED THE ONLY GREAT ADVANCE ON PLANE TABLE SURVEYING (USED BY THE ROMANS) OTHER THAN THE INVENTION OF AN ACCURATE THEODOLITE/TRANSIT.

point.[45] Surveyor General Deville began assigning men to adopt and modify this method while the survey was still on the prairies. Born in France, Deville spent his young adulthood in the French navy in charge of hydrographic surveys. In that capacity, he learned contemporary techniques of surveying and astronomic positioning before immigrating to Canada in 1874. By that time, as well, he counted Laussedat a "warm friend."[46] A DLS surveyor by 1877, he rose to become inspector of the DLS in 1881, and then surveyor general of Canada in 1885. Knowing that the mountains could not be efficiently surveyed with traditional methods, Deville consulted Laussedat's writings on photogrammetry and decided to put the practice to the test in the mountains. In 1886, he established a laboratory in Ottawa in which he conducted several experiments with telescopes, field transits, and astronomical instruments. From these he built the first Stereo-Planigraph, and combined it with a selection and adaptation of specialized survey instruments that he hoped would withstand the rigours of terrain and challenges of surveying that the Rockies presented. In particular, he improved on Laussedat's ideas by designing the short tripod used by surveyors with their transits and cameras. Thereafter, Deville initiated a photographic survey of the Railway Belt, that is, the Main Range of the Rocky Mountains adjacent to the CPR line, where the extension of subdivisions would require special treatment.

From 1887 to 1892, Deville assigned the work of phototopographic mountain surveys to James Joseph McArthur (1856–1925). In a single field season alone (1891), McArthur climbed an astounding forty-three peaks. In 1886, with William Stewart Drewry (1859–1939), DLS, he made photographs of the areas he surveyed but these were used only to illustrate his survey report that year. The beginning of phototopographic alpine surveying in Canada came the next summer. For the first time, McArthur and Drewry took full 360° panoramic sets of photographs as well as readings to determine the exact location of the stations from which McArthur made his photographs and triangulations. On 9 September 1887, with assistant T. Riley and their camera gear, but without even ropes let alone other climbing aids, McArthur climbed Mt Stephen (10,495 ft [3,199 m]), which looms over the hamlet of Field, British Columbia. This marked the first ascent by a white man of a peak in Canada higher than 10,000 ft.[47]

Drewry and McArthur (Wheeler would count McArthur the foremost early climber when the *Canadian Alpine Journal* began publishing in

Photograph of James
Joseph McArthur and
William Stewart Drewry:
"Topographer and
Assistant showing the
method of carrying
photographic
equipment.—1887."
21 x 26 cm. Courtesy
Library and Archives
Canada PA–023141.

1907,[48]) sent the negatives from this work back to Horatio Nelson Topley at the Photographic Division, the central laboratory organized in Ottawa by Deville to ensure consistency in and limit the costs of the photographic developing of work by the Surveys Branch. Once McArthur had the prints back, he used the principles of perspective to calculate topographical details between the triangulation stations visible in the photographs. Then he and Drewry could render topographical maps with annotated mountain elevations, but with no real contour lines to define the shapes, pitches, or other characteristics of individual mountains.[49] Meanwhile, using McArthur's surveying skills to test different photogrammetric theories, Deville set out to determine the most efficient method of surveying mountainous areas. In 1889 he published his findings in the book, *Photographic Surveying*, which attracted international acclaim, in part because, as Wheeler would attest in 1920, "the method has been applied in many directions and for various purposes where the highly accentuated contours of the terrain rendered the application of ordinary methods of survey too laborious and expensive, or else altogether impracticable."[50]

On 23 June 1887, Parliament passed its *Rocky Mountains Park Act*, which brought into being a smaller version of what we know today as Banff National Park.[51] It measured about 260 mi² (675 km²) in a fiercely rigid rectangle that imposed itself ponderously on the irregular terrain from just west of the Banff town site in a northeasterly direction, taking in Lake Minnewanka and extending out beyond the Front Range into the Foothills. In this year, as well, the Technical Branch of the Department of the Interior became the Topographical Surveys Branch. It dedicated itself to the topography of the eleven million acres of the Railway Belt, a looping, winding strip of the country extending right across British Columbia from the hamlet of Field to the Pacific Ocean, measuring twenty miles on either side of the CPR line. Although photogrammetric techniques had steadily improved over their first few years of use, the Railway Belt presented specific challenges that could be resolved only partially through the use of photogrammetry. Over six thousand separate calculations were needed to establish the theoretic position of the twenty mile (thirty-two kilometre) boundary on either side of the line. Also needed were multiple summers' worth of fieldwork, including photography, astronomical and plane-table readings, and the building of cairns or other markers at the limit of the allocated land. Drewry's work on the belt, up the Bow River and over the Great Divide as far as the Beaverfoot River, extended until 1892 and involved a good deal of exploration as well as surveying. In that year, he was transferred to the Alaska-Yukon boundary survey.[52]

As topographic features were not deemed necessary information on a map meant to control the sale and administration of Crown lands for homesteads, the grid system of the prairies could take the form of a plane survey, a type in which the mean surface of the earth is considered as a plane. The map that results from it does not represent depressions and elevations of terrain. By contrast, geodetic surveying, which takes into account the shape of the earth and aims to render topography in three dimensions, identifies land by a more complex method. It was such a type of surveying that the government wanted and Deville assigned for the mountains. But even geodetic methods of survey would not have met the challenge, except perhaps in the wider valleys, and would cost a young nation a prohibitive fortune. Photogrammetry was the key to the surveying and mapping of the Cordillera. No wonder, then, that Wheeler spent five seasons at the end of the century in the Foothills honing his skills as a phototopographer and modifying the methods set out in Deville's book. Painstaking efforts yielded only indifferent results, however.[53]

That is not to say that the grid survey's skeleton—meridians and base lines—was not extended into the mountains at all. Both coal and railway building required surveying in the Bow River valley as early as 1884, and in 1886, as McArthur made his tentative start with photogrammetry, Otto Klotz, DLS, ran a declination survey as far west as Revelstoke, British Columbia. Work farther north had to wait until the next century, however, chiefly because all mountain surveying in the southern parts of Alberta and British Columbia was put on hold until the needs of the gold rush and boundary dispute were met up north. Thereafter, in 1904, Arthur St Cyr (1862–1923), DLS, surveyed the upper Athabasca River valley. He completed the survey of the sixth meridian along 118°W Long. in 1904. Four years later, Albert Howard Hawkins (1862–1950), another DLS surveyor of northern British Columbia and the Yukon at the time of the gold rush, extended the thirteenth base line west as far as its intersection with the sixth meridian, just beyond the site of Jasper House II, opposite Roche Miette on the Athabasca River.[54] So townships had been established for nearly a decade in the eastern portions of what became Jasper National Park well before Morrison Parsons Bridgland arrived to

PHOTOGRAMMETRY
WAS THE KEY TO
THE SURVEYING
AND MAPPING OF
THE CORDILLERA.

conduct his phototopographical work in the summer of 1915. That was his fourteenth year of fieldwork pertaining to the mountains. Back in 1901 Wheeler had been assigned by Deville to survey the Selkirk Mountains and the glacial water sources there. A year later, Bridgland came west and joined Wheeler's phototopographical survey.

By the time Bridgland arrived, the Athabasca valley had entertained further survey work. The advent of railways had spawned a hunt for coal throughout the west. In 1908 the coal seams that Sir James Hector had first noted in 1859 at the base of Roche Miette, as well as across the Athabasca River at Bedson Ridge, had attracted more than passing interest. In fact, both in 1898 and in 1906 geological surveyors had also noted the mineral potential in the valley. Indeed, thanks to Hector, consideration of putting the transcontinental line through Yellowhead Pass had always involved the supposed proximity of coal for train loco-motives. Prompted by Hector's report, Alfred Richard Cecil Selwyn (1824–1902), the successor to Logan as the second director of the Geological Survey of Canada, had deputed himself to examine the region in the fall of 1871. However, although he reached Yellowhead Pass from the British Columbia side, a failure to rendezvous with a supply party prevented his continuing east into the Miette and Athabasca valleys. With the decision to opt for a more southerly railway route, the pass lost its interest to geological surveyors until the very end of the century. In 1898, when Bridgland was in the second year of his undergraduate degree at the University of Toronto, James McEvoy (1862–1935) travelled through it compiling an extremely detailed report of its natural resources, including the limestone beds at Roche Miette. Called a "minor classic of its genre" and "an excellent report," it opened the eyes of those harbouring or developing commercial perspectives on the West.[55] Donaldson Bogart Dowling (1858–1925) was the next reporter on the valley's mineral potential. In 1906, when, thanks to McEvoy's estimate, the Yellowhead was being already regarded as the next pass for a railway, Dowling was assigned to deter-mine the quality of coal available near or in the pass. Over the course of the remaining years of the decade, Dowling determined that two fields were economically viable: one near the modern village of Brule, Alberta, on Brûlé Lake, and the other at the eastern foot of Roche Miette.[56]

Even before Dowling had completed his report, prospectors Frank Villeneuve and Alfred Lamoureux staked a claim at Miette in 1908, and

THE ADVENT OF RAILWAYS HAD SPAWNED A HUNT FOR COAL THROUGHOUT THE WEST.

Ernest Percy Brown, surveying for the Grand Trunk Pacific Railway, 1911.
Courtesy Glenbow Archives NA–3304–13.

within two years Jasper Park Collieries was operating at the base of the mountain.[57] Its workers erected the hamlet of Pocahontas, or "Poco," after a successful coal mining town in Virginia (not because anyone appreciated the valley's Native population). Charles Andrew Grassie (1882–1950), DLS, began a grid survey of the site in 1910. The next year, Jean Leon Côté (1867–1924), DLS, invested in the mine and transformed the operation into an initial bonanza for himself and partner

Photograph by M.P. Bridgland of the Athabasca River valley from Roche Miette
showing Pocahontas, Alberta, Stn 83, no 678, direction north, 1915.
Courtesy Jasper National Park.

Frank Smith. The mine could supply locomotives directly from the
spur that passed under its tipple, thereby saving the railway the cost of
bringing coal from as far away as the Crowsnest Pass and Pennsylvania.
In fact, its bituminous coal would power Grand Trunk Pacific Railway
(GTPR) locomotives between Winnipeg and Smithers, British

Detail of 7a shows the hamlet of Pocahontas, Alberta, (*right*) in the context of the Athabasca River valley.

Further detail of 7a shows the mine and GTPR railway at Pocahontas, Alberta.

GTPR Station at Pocahontas, Alberta, c.1915. Courtesy Glenbow Archives NA–2062–2.

Columbia. The investment seemed a sure bet, but by July 1920, and for
a host of contributing reasons, the mine had shut and Côté was left
with nothing more or less than an impressive debt.[58] A similar early
craze for limestone with the advent of cement as a building construction material prompted twenty-three applications for leases by the late
autumn of 1911, but works begun upstream from Pocahontas might
have furnished only a single boxcar of burnt lime, destined for the
construction of the Hotel Macdonald in Edmonton, before fizzling
out when the GTPR's track was taken up during the war for use in
Europe.[59] Roche Miette was once again left in peace, although trains
(trains with electric lights, no less) were passing beneath it.

In the employ of the DLS in the summer of 1911, Philadelphia-born
alpinist Mary Schäffer (1861–1939) worked for D.B. Dowling in
compiling a DLS survey of Maligne Lake, which she had visited first in
the summer of 1908, and which was first seen by a surveyor in 1875,
when Henry A.F. MacLeod, a member of the CPR Survey under
Sandford Fleming, "was so wearied from the journey [up to the lake]
that he named it Sorefoot," and dismissed it as a viable railway route.[60]

Miette Hot Springs c.1920. Miners built the log pool (*foreground*) and shelters during a strike in the summer of 1919, four years after Bridgland's survey. Courtesy Glenbow Archives NA–3934–14.

Schäffer had found it a superb treasure in 1908, and probably counted herself fortunate to be paid to visit it again. But the detailed, tedious, careful work required of a surveyor wrenched her out of the role she normally assigned herself of pleasure-seeking in alpine realms. Her consolation was that Dowling approved of her work and suggested that she "send both the measurements and her map, marked with the names she had given various features, to the Canadian Geographical Board in Ottawa."[61] Her names were adopted, and mountains around the lake, such as Unwin, Mary Vaux, Warren, Paul, and Samson, thus bear the impress of her presence today.

In the same summer, when the GTPR was reaching into the Athabasca valley and Jasper Park, George Henry Herriot (1888–1966), DLS, disembarked at Pocahontas charged with the task of surveying the mouth of Fiddle River for a town site, in advance of a building boom that was anticipated as a result of the erection of a luxury hotel up the Fiddle River at what today are known as Miette Hot Springs. The site for the hotel required surveying, as well. The craze for such springs had catapulted Banff into prominence a quarter-century earlier, and those in Jasper Park were deemed to be as promising as, if not superior to, the Cave and Basin springs on the slope of the Bow River valley. Herriot also submitted his suggestion for the route of a scenic road nineteen km up the river to the springs. "In such a valley," writes Judy Larmour,

> it would prove impossible to reach the corners of all quarter sections, so Herriot was instructed to indicate them with witness monuments. Then he began the contour survey, using stadia measurements, a more convenient method to establish distances and elevations than running chained traverses in such mountainous terrain.[62]

The hotel was never built and the town never materialized at the mouth of Fiddle River, but valuable experience was gained in attempts to survey mountainous terrain without the use of a camera. The trail from Pocahontas up to the hot springs, meanwhile, was laid out in 1911.[63]

Not long after his arrival on the job in March 1913, Col. Samuel Maynard Rogers (c.1869–1940), the first resident superintendent of Jasper Park, issued orders for the survey and construction of a permanent road from Fitzhugh, the name of the divisional point on the GTPR's line, up to Pyramid Lake, northeast of the town site.[64] Later in 1913 Fitzhugh was re-named Jasper. The town site was surveyed in competent grid fashion but the surveyor general was unimpressed by the lack of imagination in town planning that it represented:

> Deville saw national park townsites as perfect places in which to implement the foremost in town planning concepts. Such towns were government-regulated, non-industrial, and dedicated to the pursuit of an idyllic lifestyle. ...When asked to comment on the plan [approved by the Dominion Parks Branch (est. 1911)], Deville disdainfully replied that it "is just a common gridiron pattern of the real estate man, made to face on 95 acres of railway yards. It is devoid of any characteristic or attractive feature and ignores every principle of town planning."[65]

The DLS and the Dominion Parks Branch decided that forestry (then, like coal mining, permitted in parks) and the expectation of tourism would benefit from accurate maps. In October, Hugh

Matheson (1879–1959), DLS, began the work upstream of the town site and then downstream of it. The onset of winter did not deter him; he even continued another month after a fire in the hotel in Jasper where he was staying consumed many of his possessions in early December. But he was back at it in the spring of 1814, continuing "the topographical survey at elevations that allowed for the use of transit and stadia rod and a plane table for plotting." This experience prompted him to suggest to Surveyor-General Deville that "if the topographical work at Jasper was to be extended into the mountains,...the surveying camera could be used very effectively."[66] Indeed, it would be, the very next summer, by the surveyor who had come to perfect its use in mountainous terrain: Morrison Parsons Bridgland. ✳

✳ *Notes* ✳

1 Elizabeth Boone Hill, "Maps of Territory, History, and Community in Aztec Mexico," *Cartographic Encounters: Perspectives on Native American Map-making and Map Use*, ed. G. Malcolm Lewis (Chicago: University of Chicago Press, 1998), 113–14.

2 Quoted by G. Malcolm Lewis, "Frontier Encounters in the Field: 1511–1925," *Cartographic Encounters*, 18.

3 Lewis, 14–15.

4 Quoted in John Noble Wilford, *The Map-makers* (1981), rev. ed. (New York: Alfred A. Knopf, 2000), 206.

5 Wilford, *The Map-makers*, 217.

6 Quoted in D'Arcy Jenish, *Epic Wanderer: David Thompson and the Mapping of the Canadian West* ([Toronto]: Doubleday Canada, 2003), 212.

7 Quoted by Irene Spry, ed., *The Papers of the Palliser Expedition, 1857–1860* (Toronto: Champlain Society, 1968), xlvii.

8 Wilford, *The Map-makers*, 240.

9 Æmilius Simpson, "Journal of a Voyage across the Continent of North America in 1826," HBCA, PAM B.223/a/3, 50 fols.

10 Simpson, "Journal," 33r.

11 Simpson, "Journal," 8 Oct. 1826, 33r.

12 See William H. Goetzmann, *Exploration and Empire: The Explorer and the Scientist in the Winning of the American West*, (1966; New York: Norton, 1978).

13 *The Journals, detailed Reports, and Observations relative to the Exploration, by Captain Palliser, of that Portion of British North America, which, in Latitude, lies between the British Boundary*

Line and the Height of Land or Watershed of the northern or frozen Ocean respectively, and in Longitude, between the western shore of Lake Superior and the Pacific Ocean during the years 1857, 1858, 1859, and 1860; presented to both Houses of Parliament by command of Her Majesty, 19th May, 1863 (London: Printed by G.E. Eyre & W. Spottiswoode, 1863).

14 Spry, ed., Papers of the Palliser Expedition, 368, 374. Spry's note reads, "With an elevation of 7,599 feet [7,600 according to today's sources] it is 4,312 feet higher than the lakes above Jasper House, which are 3,287 feet above sea level" (374n). By "lakes," she means a widening of the Athabasca River now known as Jasper Lake, as well as Talbot and Edna lakes.

15 The St Paul, Minneapolis and Manitoba Railway "had net earnings of between $1 and $2 million in each of its first two full years of operation, and came to be seen as one of the great investment coups of the era" (Gerald Friesen, The Canadian Prairies: A History [Toronto: University of Toronto Press, 1987], 175).

16 A.C. Roberts, "The Surveys in the Red River Settlement in 1869," supplement to Canadian Surveyor 24 (June 1970): 238.

17 Friesen, The Canadian Prairies, 120, 122.

18 Robert B. McKercher and Bertram Wolfe, Understanding Western Canada's Dominion Land Survey System (Saskatoon: University of Saskatchewan Extension Press, 1986), 11.

19 Andro Linklater, Measuring America: How an Untamed Wilderness Shaped the United States and Fulfilled the Promise of Democracy (New York: Walker and Co., 2002), 232.

20 Seven meridians (lines of longitude) were chosen for the Dominion survey grid system that described land in western Canada. The first was established in 1869 just west of Winnipeg. The fifth ran just west of Edmonton and through Calgary along 114°W Long., and the sixth ran just east of the townsites of Jasper, Alberta, and Revelstoke, British Columbia, along 118°W Long. (McKercher and Wolfe, Understanding). The first base line was set at the international border; thereafter, every twenty-four miles, base lines were run, so that base lines and meridians framed the structure for the surveying of the townships (the smaller parcels of land).

21 The working of this instrument is explained in section III.

22 Much information in this history is indebted to the thorough work found in Judy Larmour, Laying Down the Lines: A History of Land Surveying in Alberta ([Edmonton]: Brindle & Glass, 2005).

23 Larmour, Laying Down the Lines, 22–23.

24 Friesen, The Canadian Prairies, 182; McKercher and Wolfe, Understanding, 2, 9, 11, 26.

25 Wilford, The Map-makers, 221. See also Linklater, Measuring America, 160–62.

26 The relation between longitude and time is explained as follows:

As the sun apparently makes a complete revolution (360°) about the earth in one solar day (24 hr.), and as the longitudes of the earth range from 0° to 360°, it follows that in 1 hr. the sun apparently traverses 360/24 = 15° of longitude. The same statement applies equally well to the sidereal day and the vernal equinox [that is, the period of one complete rotation of the earth with reference to the vernal equinox]. It follows that at any instant, the *difference in local time* between two places, whether the time under consideration be sidereal, mean solar [clock time], or apparent solar [time based on the interval between two successive returns of the sun], is equal to the *difference in longitude* between the two places, expressed in hours. This relation is used to determine the difference in time when the difference in longitude between two places is known, or *vice versa*.

(Raymond E. Davis, Francis S. Foote, and Joe W. Kelly, *Surveying Theory and Practice* [1928], 5th ed. [New York: McGraw-Hill, 1966], 525–26.) An apparent solar day may differ from a mean solar day (made up of 86,400 seconds) by as much as nearly twenty-two seconds less to nearly twenty-nine seconds more. Because many of these long or short days occur in succession, the difference builds up to as much as nearly seventeen minutes early or a little over fourteen minutes late. The difference between apparent solar time and mean solar time is called the equation of time. See http://encyclopedia.thefreedictionary.com/Mean%20solar%20day.

27 *Description of & Guide to Jasper Park*, the book on which M.P. Bridgland would collaborate in 1917, is diplomatic in its expression of incredulity that the government did not follow Fleming's recommendation: "It is interesting to note that the great advantages of the Yellowhead pass ... were early recognized although not utilized for a railway until the Grand Trunk Pacific was constructed, more than a quarter of a century later" (Morrison Parsons Bridgland [and Robert Douglas], *Description of & Guide to Jasper Park* [Ottawa: Dept. of the Interior, 1917], 27).

28 George Monro Grant, *Ocean to Ocean: Sandford Fleming's Expedition through Canada in 1872. Being a Diary kept during a Journey from the Atlantic to the Pacific with the Expedition of the Engineer-in-Chief of the Canadian Pacific and Intercolonial Railways* (Toronto: James Campbell and Son; London: Sampson, Low, Marston, Low, and Searle, 1873), 226.

29 Grant, *Ocean to Ocean*, 227, 228–29.

30 Of course, the survey expedition included a photographer, Charles George Horetzky (1838–1900), on whose photographs the lithographs in *Ocean to Ocean* are apparently based. One, depicting Roche Miette from Jasper House, faces 231.

31 Daphne Sleigh, *Walter Moberly and the Northwest Passage by Rail* (Surrey, BC: Hancock House, 2003), 172.

32 Of course, this was not a grid survey, which settlers, not railways, required. On Moberly's direct defiance of Fleming's orders, see Sleigh, *Walter Moberly*, 201–02.

33 Walter Moberly, "Canadian Pacific Railway Exploratory Survey, Diary and Notes of Walter Moberly Engineer in Charge of Party S, April 19, 1872–Feburary 16, 1873," transcribed by Fred Howlett (1998), BCARS, file 334–16–12, 43; qtd by kind permission of Fred Howlett.

34 "Diary of Ashdown H. Green (transit man) C.P.R. Survey Party 'S' Jan. 31, 1872–Oct. 28, 1873," transcribed by Winnifreda Macintosh (1966), BCARS, MS-0437; [34], [35].

35 Moberly, "Canadian Pacific," 40.

36 Other sources for the railway survey of the Athabasca valley in 1872 and early 1873 include some of the photographs taken in an advance party by Horetzky and held by the Glenbow Museum. (His famous photos of Jasper House date from January 1872.) Horetzky's career is detailed in Andrew Birrell, "Fortunes of a Misfit; Charles Horetzky," *Alberta Historical Review* 19 (Winter 1971): 9–25; and W.A. Waiser, "Horetzky, Charles George," *Dictionary of Canadian Biography*, vol. XII (1891–1900), gen. eds. Frances G. Halpenny and Jean Hamelin (Toronto: University of Toronto Press, 1990), 447–48. Another member of Moberly's survey from the Columbia River over to Jasper House and beyond was R.M. Rylatt: see his *Surveying the Canadian Pacific: Memoir of a Railroad Pioneer*, fwd. by William Kittredge (Salt Lake City: University of Utah Press, 1991).

37 Esther Fraser, *The Canadian Rockies; Early Travels and Explorations* (Edmonton: Hurtig, 1969), 110.

38 Eleanor G. Luxton, *Banff, Canada's First National Park: A History and a Memory of Rocky Mountains Park* (Banff: Summerthought, 1975), 64.

39 A.O. Wheeler, M.P. Bridgland, and A.J. Campbell, "The Application of Photography to the Mapping of the Canadian Rocky Mountains," Report submitted to the Alpine Congress at Monaco by the Alpine Club of Canada, *Canadian Alpine Journal* 11 (1920): 78.

40 Don W. Thomson, *Men and Meridians: The History of Surveying and Mapping in Canada*, 3 vols. (Ottawa: Queen's Printer, 1966–1969), 2:26.

41 Further details of the history of the evolution of photogrammetry are provided by Ron Fischer, "Seeing with Metric Eyes: the Unknown Origins of Motion Capture, Part 1," *VR News* 9:2 (Mar. 2000).

42 See Robert J. Huyda, *Camera in the Interior, 1858: H.L. Hime, Photographer: The Assiniboine and Saskatchewan Exploring Expedition* (Toronto: Coach House, 1975).

43 A history of photography's role in the work of the Boundary Commission is provided in Andrew Birrell, "The North American Boundary Commission: Three Photographic Expeditions, 1872–74," *History of Photography* 20 (1996): 113–21.

44 A.O. Wheeler, M.P. Bridgland, and A.J. Campbell, "The Application of Photography," 80.

45 Triangulation uses trigonometry to determine the coordinates of latitude and longitude of a particular place from the known coordinates of two other places. It had to be used when, faced with a steep incline or a straight stretch of water that precluded the hammering of stakes into the ground, surveyors would run a line in an alternative direction, measure the angle at which it diverged from the baseline they were attempting to plot, and the length it extended from the base line. Then, measuring from its end point the angle that another line extended back to intersect with the hypothetical extension of the base line would yield

two known angles and the length of one side of the triangle thus formed, sufficient information to permit the calculation of the third angle and the other two lengths. Thereby, the base line survey could resume on the far side of the obstacle imposed by the terrain or body of water. Further explanation is provided by Davis, Foote, and Kelly, *Surveying Theory and Practice*, 388–95.

46 A.O. Wheeler, M.P. Bridgland, and A.J. Campbell, "The Application of Photography," 77.

47 Thirteen years later, when Mary Vaux summitted it, Mt Stephen, named for the CRP president, became the first such peak climbed by a woman. Riley is less well remembered than McArthur's most notable employee, Tom Wilson, the guide and horse-packer who is credited with being the first white man to see Lake Louise. (In 1882 Stoney Native Edwin Hunter took him to see his "Lake of Little Fishes.")

48 Wheeler's praise of McArthur, found in *Canadian Alpine Journal* 1 (1907): 36, is quoted in Chic Scott, *Pushing the Limits: The Story of Canadian Mountaineering* (Calgary: Rocky Mountain Books, 2000), 39.

49 Canada, Topographical Surveys Branch, Map, *Topographical Survey of the Rocky Mountains*, scale 1:40,000; triangulation by W.S. Drewry, DLS; topography by J.J. McArthur, DLS (Ottawa: Mortimer & Co. Lith., 1888–92).

50 Édouard-Gaston Deville, *Photographic Surveying, including the Elements of Descriptive Geometry and Perspective* (Ottawa: Government Printing Bureau, 1889), 2d ed. (Ottawa: Survey Office, 1895). Wheeler, Bridgland and Campbell, "The Application of Photography," 77. Wheeler's reference to the lower costs of phototopographical surveying is interesting. He argued both that the method, by furnishing photographs and not just a map, engaged a wider array of interests—"geographers, geologists and the large number of persons who are interested in mountain regions wrapped in ice and snow, who come to them annually to study many branches of natural science, not omitting artists and photographers" (81)—and that the method produced twice as effectively with no increase in manpower: "With the plane-table much of the plotting is done in the field, with the camera altogether in the office; with the plane-table you can occupy but one station at a time, but with the camera the views from both stations lie on the table before you. ... Two persons alone can conduct the camera and instrumental work at a station, and one may be a porter to help carry the instruments. It is better to have three in case of an accident and to accelerate operations" (82–83).

51 Canada, Statutes, 50–51 Victoria, Chap. 32, "Rocky Mountains Park Act, 1887," rptd. in *Documenting Canada: A History of Modern Canada in Documents*, ed. Dave De Brou and Bill Waiser (Saskatoon: Fifth House, 1992), 154–55.

52 "William Stewart Drewry, P.L.S. #14," *Early Land Surveyors of British Columbia (P.L.S. Group)*, comp. and ed. John A. Whittaker (Victoria, B.C.: Corporation of Land Surveyors of the Province of British Columbia, 1990) 46–47.

53 These years of work are well summarized in Larmour, *Laying Down the Lines*, 62–66.

54 Archives of the Surveyor General of Canada Lands, Legal Surveys Division, Earth Sciences Sector, Department of Natural Resources Canada. The fourteenth base line marks the line along which Jasper Avenue, the principal east-west street in downtown Edmonton, runs.

55 Great Plains Research Consultants, "Jasper National Park: A Social and Economic History," typescript report for Parks Canada (1985), 84; John A. Whittaker, ed., *Early Land Surveyors*, 90; James McEvoy, *Report on the Geology and Natural Resources of the Country traversed by the Yellow Head Pass Route from Edmonton to Tête Jaune Cache, comprising Portions of Alberta and British Columbia*, Geological Survey of Canada, Annual Report, vol. 11, pt D (Ottawa: S.E. Dawson, 1900).

56 Canada, Sessional Papers, Department of Mines, *Annual Report for 1910* (Ottawa, 1911).

57 The name, Jasper Park Collieries, is likely to surprise those who have not heard it before, given its apparently contradictory celebration of mining activity in a national park. In fact, however, the idea that parks must pay for themselves by harvesting their own natural resources held sway in national parks policy until 1930, when the *National Parks Act* at last prohibited industrial activity of any kind. The standard analysis of the early parks policy's emphatic promotion of resource extraction is R.C. Brown, "The Doctrine of Usefulness: Natural Resource and National Park Policy in Canada, 1887–1914," *The Canadian National Parks: Today and Tomorrow*, eds. J.G. Nelson and R.C. Scace (Calgary: University of Calgary, 1968; Montreal: Harvest House, 1969), 46–62.

58 Great Plains Research Consultants, "Jasper," 97–99; Larmour, *Laying Down the Lines*, 143-44.

59 Great Plains Research Consultants, "Jasper," 92–93.

60 Janice Sanford Beck, *No Ordinary Woman: The Story of Mary Schäffer Warren* (Calgary: Rocky Mountain Books, 2001), 65. Although MacLeod dismissed the lake, his visit was commemorated in 1902 by Arthur Philemon (A.P.) Coleman (1852–1939), when he named a mountain for him off the lake's southern end, at the head of Coronet Creek.

61 Beck, *No Ordinary Woman*, 102. See also E.J. Hart, ed., *A Hunter of Peace: Mary T.S. Schäffer's Old Indian Trails of the Canadian Rockies, with her heretofore unpublished Account 1911 Expedition to Maligne Lake* (Banff: Whyte Museum of the Canadian Rockies, 1980).

62 Larmour, *Laying Down the Lines*, 133. The stadia method uses sight lines rather than chain links. It is predicated on the idea that the corresponding sides of similar triangles bear a proportional relationship to each other. A theodolite (transit telescope) for stadia has two more horizontal cross hairs than normal. These are called stadia hairs. Most shots in a stadia survey are inclined because of varying elevations in the terrain, but surveyors looking through the transit at a rod held vertically in the distance—across a valley, say—can read the intercept, correct for the vertical angle thanks to the stadia hairs, and thereby calculate a true horizontal distance. Once the horizontal distance is known, they can also calculate the vertical distance involved. Thus, the length of the terrain's slope can be

determined. Surveyors direct their line of sight towards the rod as it is held vertically in the distance. Aware that the interval between the two stadia hairs measures 1/100 of the distance to the rod, they know that there will be 1 ft of intercept (the part of the rod seen between the upper and lower stadia hairs of a transit telescope) for every 100 feet of distance to the rod. With this information they can proceed to plot the traverse, which amounts to a series of straight lines run out from a known point (called a contol, or control point), with the distance, and the vertical and horizontal angles measured at each turning point. When combined with the triangulation work (often using the same control points), the stadia survey provides the framework for a subsequent topographical map. Where possible, surveyors determine where one traverse line intercepts an already established reference point. While not as accurate as the chaining of distances, a stadia survey is much faster in uneven terrain, and sufficiently precise for topographical work.

63   Great Plains Research Consultants, "Jasper," 220.

64   Great Plains Research Consultants, "Jasper," 25, 26. Earl Hopkins Fitzhugh, a vice-president of the GTPR, bestowed his name on the divisional point and thus the community in 1911, but the survey of the town site resulted in the change of name to Jasper.

65   Great Plains Research Consultants, "Jasper," 64, quoting LAC, RG 84, vol. 523, file J19-1, Deville to James Bernard Harkin, 12 Apr. 1913.

66   Larmour, *Laying Down the Lines*, 136, 137.

{ IN NOVEMBER 1905, A GROUP OF ENGINEERS, SCIENTISTS, AND RECREATIONAL ALPINISTS PONDERED HOW THEY MIGHT PROMOTE MOUNTAINEERING AS A NATIONAL SPORT.

BRIDGLAND'S LIFE AND TIMES
1878–1914

✳

Bridgland's Roots

Morrison Parsons Bridgland was a fourth-generation Canadian. In 1816, his great-grandparents, James William Bridgland and Eleanor Beaton Bridgland, journeyed with their five children to York (Toronto) from Kent, England. Soon after their arrival, James was able to secure the position of keeper of the Court of King's Bench, a division of Britain's High Court of Justice. He retired from his uneventful work in 1828, the year he moved his family to a farm that he purchased at Downsview in the township of North York. As for our surveyor's other set of great-grandparents, Jacob Parsons, a tailor in Laverton, England, before he enlisted as a soldier in the British army, came to Canada when he was awarded a land grant just a few plots south of the Bridgland property. Like the Bridglands, Jacob emigrated with his wife and young children. The Parsons and Bridgland families did not know each other before their emigration but were likely brought together in the new country by their common religious affiliation, Wesleyan Methodism.

WHILE IT MIGHT BE AN
OVERSTATEMENT TO
SAY THAT M.P. CAME
FROM A FAMILY WITH
A LONG TRADITION OF
SURVEYORS, HIS
GREAT-UNCLE PAVED
THE WAY BY DECIDING
WHEN HE WAS TWENTY-
FIVE TO BECOME A
SURVEYOR.

As the superintendent of education for Canada West and later Ontario from 1844 to 1872,[1] Adolphus Egerton Ryerson (1803–1882), the leading Methodist of his day, was able to institute policies that were instrumental in establishing the common school (elementary school), normal school (teacher-training school), and public high school, from which all of the Bridgland and Parsons children reaped the benefits. Ryerson's work also influenced the development of denominationally-run universities. In 1841, he became the founding principal of Victoria College, the pre-eminent Methodist college of its day, which was to become M.P.'s *alma mater*, and had been attended by his great-uncle, James William Bridgland II (1817–1880), the first child born to James and Eleanor Bridgland after their arrival in Canada.[2]

While it might be an overstatement to say that M.P. came from a family with a long tradition of surveyors, his great-uncle paved the way by deciding when he was twenty-five to become a surveyor. He apprenticed under Col. John Stoughton Dennis, who, as we have seen, held several positions of distinction in the bodies governing land and surveys throughout his career. Although James had no direct involvement with the Rebellion, his work as the inspector of colonization roads brought him very close to the centre of the dispute while he surveyed and directed the building of a road from Fort William to the Red River Settlement. Throughout his career he directed the surveying, planning, and construction of major roads to the north and west of Lake Superior, designed to encourage and expedite the settlement of Canada's interior. He made his more important early surveys in the areas around lakes Simcoe and Couchiching on land adjoining the

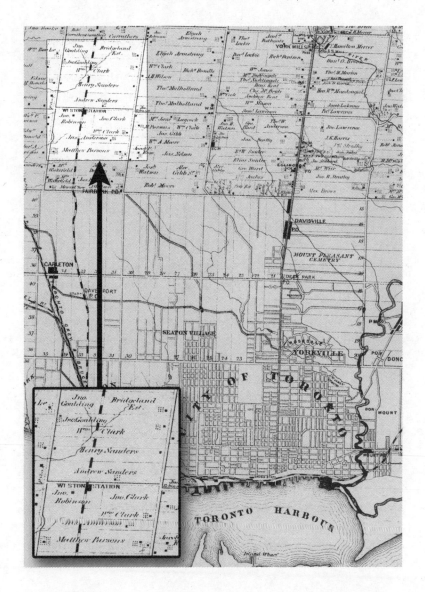

Map of Toronto and environs in 1878, the year of M.P. Bridgland's birth.
Detail showing the farms of the Bridgland and Parsons families. *Historical Atlas
of York County, Ontario* (1878), 39.

Northern Railway, and slightly farther north, into the Muskoka Lakes
region of Canada West.[3]

After 1860, James Bridgland's duties lay in the inspection and overseeing of colonization roads between Ottawa and Georgian Bay,

roughly along the line that Ontario's Highway 60 follows today. The
death of the superintendent of colonization roads in 1864 increased
James' responsibilities, effectively making him both the inspector and
the director of road surveys on the north shore of Lake Superior. He
was twice married, first to Marie Dennis, daughter of Col. Dennis, and
later to Martha Ann Jones, the daughter of a Wesleyan Methodist
minister. He had several daughters but no sons, so his descendants did
not carry on the family name. Rather, his brother Clarke named one of
his twelve children James William Bridgland III. While James was away
surveying, Clarke lived and worked on the family farm at Downsview.
He married Elizabeth Johnson and with her had five sons and seven
daughters.

The Parsons family was growing also, as were its land holdings, with
lots of various sizes in the townships of Etobicoke, North York,
Vaughan, and Markham.

Jacob Parsons' two sons, Matthew and Clarke, had taken on the
responsibilities of running the farm on the original land grant at
Fairbank.[4] In 1841, at the age of twenty-five, Matthew Parsons married
Elizabeth Sophie MacKay, with whom he had at least two children.
Little is known of Matthew's life except that he continued to inhabit
the farmstead at Fairbank. Eventually, the two families' histories
became permanently entwined when Matthew Parsons' children
married two of Clarke Bridgland's: William Abbott Parsons married
Annie Jane Bridgland, and his sister Hannah Matilda Parsons married
Annie's brother, James William Bridgland III. These two couples would
figure prominently in our surveyor's life. When Matthew Parsons
retired from farming, he left his property to these two couples. They
each had a house at opposite ends of the lot, which was divided in
almost perfect halves by the line of the Northern Railway of Canada.[5]

✳ *Bridgland's Early Years* ✳

ON 20 DECEMBER 1878, William and Annie Parsons welcomed
into their family their sixth child, whom they named Alfred Morrison
Parsons. He did not stay long in the Parsons homestead with his four
brothers and one sister: unwell during her pregnancy, his mother never
regained her health. James and Hannah, her brother and sister-in-law,
took the infant to their home to care for him, having no children of
their own. The Bridglands grew fond of their nephew, even joking with

his parents that they did not want to give him back. When within a couple of years Morrison's mother's died, the two families agreed that James and Hannah would formally adopt the young boy. When they did so, he was given the name Morrison Parsons Bridgland.

M.P. thus grew up knowing two family homes, one at each end of the 200-acre Parsons estate, located between modern Toronto's Keele and Dufferin streets, about one-half km north of Eglinton Avenue. He lived with his Uncle James and Aunt Hannah in the red brick Bridgland house on the eastern boundary of the property. On the western boundary, he lived with his five Parsons siblings in the white-stucco farm house where he was born. When it came time for high school, he had a fair distance to go. The one he attended from 1893 to 1897 opened in 1892 as the Toronto Junction Collegiate Institute. Still a fully-functioning high school, it now bears the name of Humberside Collegiate, and is located in the High Park area of western Toronto. So from north of Eglinton Avenue down to Bloor Street and then west made for a long haul back and forth every day. Less than ten percent of students graduated when the public school system was first established, not because most students failed but because many opted to take up one of a variety of employment opportunities. For his part, M.P. completed all four forms, or grades, taking classes in Latin, Greek, Algebra, Geometry, Arithmetic, Botany, and three histories: Canadian, English, and Ancient.

In 1897, M.P. entered the B.A. Program in Maths and Physics at Victoria University, Toronto. If not beforehand, during his university years, he became known as Morris to his friends and as M.P. to his casual acquaintances. His record of student involvement shows interests not only in athletics but also in photography and in experiments that studied the effects of x-rays and sparks from cathode ray tubes on black and white glass plate negatives, experiments the likes of which would produce television technology three decades later. He published a paper on his findings in *Acta Victoriana*, the student magazine.[6]

In his freshman year at Victoria, *Acta Victoriana*, which routinely teased all the incoming students in short introductory write-ups, alluded to M.P.'s evangelical and rural upbringing:

Morrison Parsons Bridgeland [*sic*] comes from Fairbank Sunday School. Says he feels rather homesick—a good sign in a Freshman. Early in life "he lisped in numbers, for the numbers came" and so he is now taking Honor Math. The next time "City *vs.* Country Life" is resurrected, he can be put down for the country.[7]

From the year of his birth up until his university years, Bridgland's country had undergone vast changes. In particular, the profession of engineering was dramatically altering the face of the land. In 1871, British Columbia's willingness to join Confederation on the promise of a railway testified to a remarkable faith by Canada's two million people that such a project would prove economically beneficial to the Dominion. While the capital needed to accomplish such a project appeared beyond the young country's reach, so did the necessary knowledge. In that same year, fewer than five hundred engineers lived in all of Canada, and no engineering school existed. How would Canada be able to survey the route let alone complete the construction? Proposals circulated for different ways of educating the needed engineers. The apprenticeship system, which had been adopted in Britain, was not feasible in a land with so few qualified experts, and so in 1873 the Ontario government passed an act to establish the Ontario School of Practical Sciences (SPS) at the University of Toronto.[8]

After receiving his Honors Maths degree in 1901, M.P. enrolled in the second year of the engineering course at the SPS.[9] You would think that engineering was already on his mind, but apparently not. His year book put him down for a future as a teacher:

> Maurice [*sic*] P. Bridgland exemplifies the old adage, "Genius is but an infinite capacity for hard work." M.P. has proved himself almost more than a star in practical physics. You should see some of his photos of electric sparks! At Association football and at alley he is hard to beat, and is what one calls in college vernacular "a good square head." May he have all success in his chosen vocation of teaching.[10]

Although M.P. never faced students in a classroom, his education and career would make him a teacher of many. It is not difficult to trace this aspect of his personality in his work as a surveyor and handler of men. Similarly, his infinite capacity for hard work is readily identifiable in that summer's survey in Jasper in 1915. No wonder he was so prolific as a surveyor and alpinist. No wonder either that a young assistant in 1918

ALTHOUGH M.P. NEVER FACED STUDENTS IN A CLASSROOM, HIS EDUCATION AND CAREER WOULD MAKE HIM A TEACHER OF MANY.

Graduating Class, Victoria University, 1901. M.P. Bridgland in the back row, *far right*.
Acta Victoriana vol. 24, no 8 (1901), 425.

felt relief at being taken off M.P.'s party and assigned to a less energetic surveyor's. If anyone embodied Scottish philosopher Thomas Carlyle's famous dictum about genius being a matter of hard work, it was Morrison Parsons Bridgland.

✳ *A Career of Adventure* ✳

EARLY IN 1902, at the end of his first term at the SPS, Bridgland was encouraged to apply for a summer position as an assistant on a Dominion Lands Survey team. The encouragement came from one of his professors, William James Loudon (1860–1951).[11] The job paid a reasonable salary, two dollars per day, but the chief attraction for M.P., who had always been a keen athlete, was likely the opportunity to spend a summer climbing, photographing, and camping in the Canadian Rockies. As his classmate, teammate, and long-time friend Charles Bruce Sissons (1879–1965) once noted, Morris "found much deep, if quiet, satisfaction in the contemplation of nature in her sterner as well as her more benign aspects."[12] Country over city, indeed.

Moreover, the prospect of a first-ever trip west must have sounded great to an Ontarion.

The first decade of the century proved to be heady times for young men. Nature beckoned. The Klondike Gold Rush had ended the previous century with a spirit of dizzying exhilaration, but there were less crazed and precarious ways of immersing oneself in the Back of Beyond. Charles G.D. Roberts, Ernest Seton, and Rudyard Kipling had begun a vogue for animal stories in the early 1890s; Jack London would publish his story, "Call of the Wild" in *The Saturday Evening Post* in June and July 1903; and four years later Robert Service would publish his two entrancing volumes of poetry, *Songs of a Sourdough* and *The Spell of the Yukon*. Into the world came his "Call of the Wild" in stirring versified form:

> Have you gazed on naked grandeur where there's nothing else to gaze on,
> Set pieces and drop-curtain scenes galore,
> Big mountains heaved to heaven, which the blinding sunsets blazon,
> Black canyons where the rapids rip and roar?
> Have you swept the visioned valley with the green stream streaking through it,
> Searched the Vastness for a something you have lost?
> Have you strung your soul to silence? Then for God's sake go and do it;
> Hear the challenge, learn the lesson, pay the cost.
>
> Have you wandered in the wilderness, the sage-brush desolation,
> The bunch-grass levels where the cattle graze?
> Have you whistled bits of rag-time at the end of all creation,
> And learned to know the desert's little ways?
> Have you camped upon the foothills, have you galloped o'er the ranges,
> Have you roamed the arid sun-lands through and through?
> Have you chummed up with the mesa? Do you know its moods and changes?
> Then listen to the Wild—it's calling you.
>
> Have you known the Great White Silence, not a snow-gemmed twig aquiver?
> (Eternal truths that shame our soothing lies.)
> Have you broken trail on snowshoes? mushed your huskies up the river,
> Dared the unknown, led the way, and clutched the prize?
> Have you marked the map's void spaces, mingled with the mongrel races,
> Felt the savage strength of brute in every thew?
> And though grim as hell the worst is, can you round it off with curses?
> Then hearken to the Wild—it's wanting you.

Have you suffered, starved and triumphed, groveled down, yet grasped
    at glory,
Grown bigger in the bigness of the whole?
"Done things" just for the doing, letting babblers tell the story,
Seeing through the nice veneer the naked soul?
Have you seen God in His splendors, heard the text that nature renders?
(You'll never hear it in the family pew.)
The simple things, the true things, the silent men who do things—
Then listen to the Wild—it's calling you.

They have cradled you in custom, they have primed you with their preaching,
They have soaked you in convention through and through;
They have put you in a showcase; you're a credit to their teaching—
But can't you hear the Wild?—it's calling you.
Let us probe the silent places, let us seek what luck betide us;
Let us journey to a lonely land I know.
There's a whisper on the night-wind, there's a star agleam to guide us,
And the Wild is calling, calling … let us go.[13]

Mountain surveying was being pitched as a career of adventure for the adventuresome. And why not? Although sketch maps existed, the job still required the breaking of new trails through unfrequented passes. The Department of the Interior recruited young men in their early twenties, giving preference to those who had taken courses in engineering and maths beyond the undergraduate level, who could plot, triangulate, and calculate well, and who could or at least were willing to try to withstand the harsh conditions of mountain travel. Moreover, because during this era surveyors enjoyed none of the lightweight items we might call "essential" outdoor gear today, the men who succeeded best at the job demonstrated generally advanced physical fitness.

MOUNTAIN SURVEYING
WAS BEING PITCHED
AS A CAREER OF
ADVENTURE FOR
THE ADVENTURESOME.

On 20 February, Morris wrote his Preliminary Examination and qualified for his articles. This marked the first step towards accreditation as a surveyor with the Dominion Lands Survey. By this point in the survey's existence, most DLS hopefuls had to sit five three-hour exams, one each in the subjects of algebra, plane geometry, plane trigonometry, spherical trigonometry, and mensuration and superficies. Since Bridgland had completed a post-secondary degree in a field related to surveying, he sat the "limited examination" and passed.[14]

His promise and now this success attracted the notice of Foothills-hardened A.O. Wheeler, who, on the recommendation of Loudon, hired M.P. as his assistant for the survey he had begun the previous season in the Selkirk Mountains. This range lies west of the Rockies and Purcells, between Golden and Revelstoke, British Columbia, on the CPR line, and includes Rogers Pass. Bridgland was a promising assistant but, with only the single term at the SPS under his belt and no field experience in surveying or mountain climbing, he no doubt made the week-long train trip to Alberta with some anxiety. Arriving in Calgary in mid-winter, he caught his first glimpse of the Rocky Mountains that would be his summer research station for the next quarter-century. He made arrangements to keep a room at a boarding house in Calgary until June, when he embarked on his first alpine adventure.

Wheeler was confident that, by training under him, Bridgland would probably acquire a practical knowledge of topographic surveying in less time than at the SPS.[15] He was probably right: DLS commissions were considered equivalent to, if more specialized than, graduate degrees in engineering. Learning mountaineering skills at the same time would present a stiff challenge, one not made any easier by the teaching philosophy of the chief of party. Wheeler had no mountaineering experience himself before he began surveying the Selkirks in 1901. Although after arriving in Canada from Ireland, he trained in northern Ontario, and spent seasons in Manitoba, Saskatchewan, and most recently the Foothills, he was just getting his feet wet in the heart of alpine terrain. He was prudent enough to realize that he should not attempt climbing on ice and snow fields without some instruction. So, to accompany him on his first high climb in 1901, he had engaged four of the Swiss guides whom the CPR had brought to the Rockies in order to develop tourism. He soon dismissed them, however, deciding that his "own men—Canadians—were before the end of the season, quite their equals; and in the bush and brush could give them points."[16] He remained friendly with several of the Swiss guides throughout his long career, but he seldom engaged their professional help.

Wheeler's bouts of bad temper were well known to those who worked for him, and so was his demanding nature. The twenty-three-year-old apprentice must have handled his boss professionally and patiently, for, year after year, Wheeler specifically requested that the surveyor general appoint Bridgland as his assistant. Not only a hard-

ARRIVING IN CALGARY
IN MID-WINTER, HE
CAUGHT HIS FIRST
GLIMPSE OF THE
ROCKY MOUNTAINS
THAT WOULD BE HIS
SUMMER RESEARCH
STATION FOR THE NEXT
QUARTER-CENTURY.

working and technically skilled engineer, he also proved to be a natural mountaineer. Within weeks of learning basic climbing skills, M.P. could climb faster and on more difficult routes than his teacher.[17] Although the two were always amiable with one another, Wheeler's biographer notes that Bridgland once responded to an impromptu visit in the field by confessing that he *"hated* to see the return of the chief" to camp.[18]

Overt animosity apparently surfaced between them only rarely but their personalities could hardly have been more different. Wheeler was a very public person, with an international reputation as a distinguished surveyor and mountaineer. His social circle had the air of snobbery about it—he railed at the DLS when it adopted an amateur's (Mary Schäffer's) proposed names for physical features around Maligne Lake[19]—and he was known to all his surveying colleagues for his ability to curse a blue streak when frustrated with his horses, the weather, or the work he was trying to accomplish. (He maintained a public persona of dignity and sophistication, however.) By contrast, Bridgland was retiring; his quiet demeanour never attracted the sort of attention on which Wheeler thrived. As well, where Wheeler played the patriarch—"The Grand Old Man of the Mountains" was his sobriquet—Morris was collegial, wanting nothing more than to climb and work in peace, and help others learn to do the same. Although in the end internationally esteemed as an authority on photographic surveying, Bridgland never garnered the same stature as Wheeler in histories of the mountain surveys, perhaps because his work spanned so many different ranges in Alberta and British Columbia. Doubtless, Wheeler's fiery as opposed to Bridgland's temperate personality also had much to do with the respective attention and neglect paid them. Still, it was just like both Wheeler and Bridgland for a multi-authored publication titled "The Application of Photography to the Mapping of the Canadian Rocky Mountains" to appear prominently under Wheeler's name. This constituted a "Report submitted to the Alpine Congress at Monaco by the Alpine Club of Canada," so international recognition attended its publication and dissemination.[20]

With M.P.'s able assistance, Wheeler successfully created a detailed topographic map of the Selkirk Range and then proceeded to several other phototopographic surveys. Some of these ventured far into unknown ranges, but most were initially conducted in high-traffic tourist areas along the CPR line. Although the government wanted a full inventory of the mountains—their topography and their resources

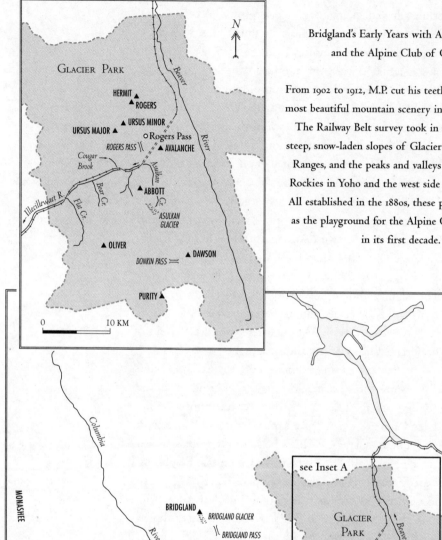

Bridgland's Early Years with A.O. Wheeler and the Alpine Club of Canada

From 1902 to 1912, M.P. cut his teeth on some of the most beautiful mountain scenery in western Canada. The Railway Belt survey took in the memorably steep, snow-laden slopes of Glacier Park's Columbia Ranges, and the peaks and valleys of the western Rockies in Yoho and the west side of Banff parks. All established in the 1880s, these parks also served as the playground for the Alpine Club of Canada in its first decade.

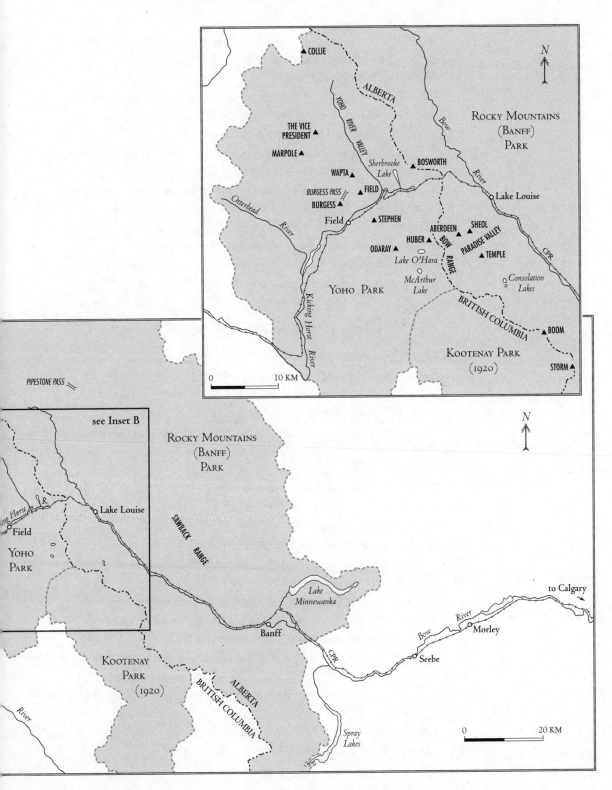

—it made most economic sense to map first the areas that held the promise of revenue from tourists, and to proceed farther afield as time and the federal budget would allow.

During his first years on the job, Bridgland spent the winter months performing some of the many tasks involved in converting the survey data into maps. The office in Ottawa, where many of the maps were actually drafted and where negatives and prints were developed of the photographic surveys, had steadily expanded in size and scope since its inception. By 1903 it employed 106 people, including eleven astronomers, nine geographers, seven darkroom operators, thirteen lithographers and printers, six draftsmen, thirty clerks to examine and compile township plans, and various other personnel. In the first few seasons of photographic surveying, the Ottawa office handled an average of three thousand photos per season. By 1910, this number had increased to over twenty thousand.[21]

IT'S QUITE ASTONISHING THAT WHEELER LEFT BRIDGLAND ON HIS OWN SO EARLY. SINK OR SWIM, FALL OR CLIMB, SEEM TO HAVE BEEN THE CHOICES HE OFFERED HIS MEN.

✳ *Bridgland's First Year in the Mountains* ✳

FOR PART OF THE SUMMER OF 1902, the men worked in two sections: Arthur, Hector (his younger brother), and Oliver (Arthur's son) in one party, and Bridgland and a few others in the other. The latter rejoined Wheeler's party at the main camp locations almost every week to replenish their provisions, but for the most part they were on their own. It's quite astonishing that Wheeler left Bridgland on his own so early. Sink or swim, fall or climb, seem to have been the choices he offered his men. Both parties had their own pack horses; the surveyors very rarely rode. The Otterhead River area presented many challenges to Morris and Charlie Sissons, especially the rough terrain and the scarcity of camp locations with sufficient pasture for the horses, or water and firewood, or even brush for themselves. It sometimes must have felt as if they worked for the horses and not the other way round. When they moved camp, plenty of axe-work faced them every few metres in order to clear a trail wide enough for the horses to pass with their bulky packs. The exasperating and shrewdest of them would balk at a pass if they thought the trail too narrow, and would obstinately refuse to proceed until the crew found a route that suited them.

The survey of the Selkirks had been commissioned by Surveyor General Deville "in order that a description and map might be prepared of this portion of the country which is much visited during the summer months by tourists and mountain climbers."[22] Partly because of uncooperative weather, Wheeler had managed to map only a small portion of the range in 1901. In this second season, he hoped to survey beyond Rogers Pass and farther up the Beaver and Illecillewaet river valleys, all within the boundaries of Glacier Park, which had been established in 1886. Motorists will readily recognize these names: the Trans-Canada Highway proceeds west up the Beaver River valley to Rogers Pass, and then down the Illecillewaet River valley to Revelstoke. Readers of Canadian poetry will doubtless call to mind Bliss Carman's lyrical "May in the Selkirks" (1929), with its allusion to Mary's expression of joy at the blessing of being made the mother of the Saviour of the world, a sense of exultation that the landscape called up for the poet more than a quarter-century after Bridgland first saw the range:

> Up the Illecillewaet and down the yellow Beaver,
> Over skyward passes where snow-peaks touch the blue.
> Shining silver rivers dropping down from Heaven,
> With the spring-call of the wilderness waking Spring anew.
>
> Far gleaming glaciers like the Gates of Glory,
> And the hosts in new green marching up the slopes,
> Organ-voiced torrents singing through the gorges,—
> Songs for the high trail and visions for our hopes.
>
> Hints of light supernal on the rocky ledges,
> Echoes of wild music from the valley floors,
> And the tall evergreens watching at the Threshold,—
> Keeping the silence of the Lord of out-of-doors.
>
> Balm out of Paradise blown across the canyons
> From the balsam-poplar buds and bronze leaves uncurled. ...
> Soul in her wonder lifts the new *Magnificat.*
> Alight with the rapture of the morning of the world.[23]

The region inspired the surveyor into print, as well. Wheeler compiled a comprehensive guidebook, history, and collection of maps in the two volumes of his *The Selkirk Range* (1905). No doubt, Bridgland's *Description of & Guide to Jasper Park* (1917) owes something to the approach taken by his predecessor, although Bridgland's is directed rather more at the tourist, Wheeler's at the alpinist.

As it was in the first year of the Selkirk survey, the crew's main challenge before the start of the 1902 field season was to transport camp supplies, provisions, and instruments into the inner reaches of their "field" of work. The region offered only two known trails: one followed Asulkan Creek; the other followed Flat Creek up to the Incomappleux River valley. On 27 July Wheeler sent the packers and the pack ponies in a freight train boxcar to the summit of Rogers Pass, where he knew there would be adequate pasture for the animals. He, Hector Wheeler, and Bridgland arrived in the pouring rain of late evening two days later, and decided to camp in the boxcar on the railway siding rather than attempt to set up their tents in wet weather.

✳ *Rocks and Hard Places* ✳

WHAT DID BRIDGLAND HAVE TO LEARN? Certainly, that good technique required knowledge of the geology of the mountains, a subject not covered at all in an Honours BA Maths degree or in the experience of growing up on a farm in southern Ontario. By knowing the types of rocks that made up each peak, how they were formed, and their relative integrity or tendencies to crumble or slide, he would learn how to choose ascendable routes up never-before-climbed peaks on his first or second try. As he acquired that knowledge, he would have his share of brushes with land slides and avalanches. One of these was recalled years later by Ley Edwards Harris (1890–1983), Bridgland's assistant and colleague for many years, but also the young man who in 1918 was relieved to be assigned to another party than Bridgland's after discovering how energetic M.P. was: "I thought I'd be glad to go with that older man," he recalled later in life, "but boy can he travel."[24] Bridgland and the survey team had just descended a mountain when they heard a loud rumbling. They took shelter below an overhang as chunks of rock flew in all directions. When the rumbling ceased, the landslide settled, and the crew rejoiced at having escaped imminent danger, Bridgland said simply, "Good thing we didn't get hit with any of that." Such was their relief that a few minutes passed before they realized that one of them had, in fact, been hit. The man in question was not seriously injured, but the large bump on his forehead sobered them all with the realization that the incident could have been fatal.[25]

Learning the geology of the mountains presented one lesson; learning how to read and respond to alpine weather presented another.

The morning of 30 July 1902 dawned fresh and clear but the fine weather did not hold. In continuous rain, the party took to the Asulkan Trail, travelling first through the forests of giant fir, hemlock, and cedar that bordered the creek, soaked by the dense undergrowth of huckleberries and punishing devil's club. Leaving the forest, they crossed a log bridge to the eastern side of the creek. Here the forests turned to sparser stands of spruce through which the men could see the dark rock face of Mt Abbott (8,051 ft [2,454 m]) looming above them. Before long, they came to the upper end of the valley, from where they could clearly see the ice tongue of Asulkan Glacier, partially buried in the boulders, mud, and other debris that formed its terminal moraine. One might suppose that the cloudy weather dispirited the young men who were making their first forays into the great mountain ranges, but Wheeler observed that "the little that could be seen was far more awe-inspiring than an unclouded view, for the heights seemed impossible of attainment and the depths unfathomable."[26]

The party stopped for a brief lunch before hiking to the summit of the pass over the ice, slush, and streams that covered the surface of the glacier. They found a cramped, makeshift shelter built by railway workers. It looked at first like a refuge of sorts, but it lacked a roof and had an uneven floor; in fact, it offered no protection from the elements. After building a small fire to refresh themselves, they decided to shoot no photographs. Only three days later did the weather offer suitable conditions for photography. The lesson to be learned about alpine weather was growing obvious: when you go into the mountains, your pack had better be filled with an ample measure of patience.

With packers, ponies, and surveyors shouldering loads of tents, blankets, instruments, and provisions, the men all worked their way single-file along Bear Creek, forging a path as they went. When they arrived at the summit of the divide at the head of Rogers Pass, they set up a second camp, from which they worked for about a week. They surveyed most of the peaks in the vicinity of Cougar Brook and Bear Creek, areas which were full of grasslands with occasional rock outcrops, and yellow dog-tooth violets on the valley's east side. As grassy meadows are a favourite habitat for bears, Wheeler recommended the names of Ursus Major (8,875 ft [2,705 m]) and Ursus Minor (9,019 ft [2,749 m]) for the surrounding peaks, but oddly gave the honour of major to the lower peak.

The men saw no bears in this vicinity. Later in the summer they did have one encounter that was too close for comfort in the area of Mt

IN CONTINUOUS RAIN, THE PARTY TOOK TO THE ASULKAN TRAIL, TRAVELLING FIRST THROUGH THE FORESTS OF GIANT FIR, HEMLOCK, AND CEDAR THAT BORDERED THE CREEK, SOAKED BY THE DENSE UNDERGROWTH OF HUCKLEBERRIES AND PUNISHING DEVIL'S CLUB.

Photograph by A.O. Wheeler of the bird's eye view of Rogers Pass, British Columbia (1909).
Courtesy Glenbow Archives NA–1869–6.

Dawson (11,122 ft [3,390 m]), on the other side of the Illecillewaet Glacier. It happened near the end of August. They noticed very large footprints along the path worn by the passage of the eight pack ponies some time earlier. Their trepidation increased as they approached camp. On reaching the supply tent, they found it torn into strips and chewed in several places. Boxes of provisions were broken and scattered all around. Two full hams, a side of bacon, a tub of butter, and many pails of jam had made a fine dinner for the hungry beast. It even left tooth marks in the leather case of one of the survey cameras, and tested the flavour of some glass plate negatives. Luckily not to his liking, these were not left in an irreparable state.

All summer long, the men pushed into the inner ranges of the mountains. As the work extended into previously unsurveyed glacial areas fraught with unpredictable crevasses, cornices, and caves, Wheeler had to resign himself to engaging the services of a Swiss guide. However, on the one occasion when such professional help would have been most beneficial, the men had to fend for themselves. It occurred in early September. Foul weather had forced them to give up surveying on Mt Dawson and to try to return to the Rogers Pass camp. To do so while enveloped in a thick fog, they had to cross the Asulkan Glacier's névé (the accumulated snow that forms the source of a glacier). The visibility was so poor that the men were forced to keep within four or five metres of one another. White walls seemed to surround them as they walked cautiously, prodding the snow with their long ice-axes before taking each step. On the occasions when an axe disappeared into emptiness, the man in the lead would prod ahead to determine how wide the gulf was; if too wide to jump, they would prod carefully and step slowly along the edge until they found a crossing place. The occasional whistle of a train engine at Rogers Pass station provided one means of orientation. Eventually, the clouds lifted and the landscape that was then revealed to them seemed comforting. As soon as they reached their destination, the weather cleared and they determined to return to their camp at Mt Dawson, but prudently hired the Swiss guide, Friedrich Michel, to accompany them back over the snow and ice.

WHITE WALLS SEEMED TO SURROUND THEM AS THEY WALKED CAUTIOUSLY, PRODDING THE SNOW WITH THEIR LONG ICE-AXES BEFORE TAKING EACH STEP.

One of the highlights of the summer was the surveyors' discovery in Donkin Pass of an old campground, which they presumed belonged to some prospector or explorer. At first they found a teapot and a frying pan, but further investigation uncovered a package of Swiss edge nails for climbing boots and several tins of preserved meat. Knowing their mountaineering history, they concluded that this site had been the camping place of Swiss mountaineer Emil Huber, prominent British mountaineer Harold Ward Topham, and his friend Henry Forster, who explored the area together a dozen years before, in 1890.[27] Arthur, Hector, and M.P. each took some souvenirs, including a tin of Armour's Corned Beef. They knew perfectly well that the meat had lain beneath the snow and sun for a long time but that evening they agreed that the corned beef would relieve the tedium of their diet of salt pork. The contents emitted no noxious odour and exhibited no obvious discolouration, so they gobbled up the treat and lived to tell the tale.

The lack of fresh foods was the most prominent drawback to mountaineering in the early years, and indeed remains today a challenge on long treks in remote regions. Then as now, dried foods were preferable to cans because of weight. Staples in 1902 included bacon, ham, rice, beans, potatoes, onion, and dried fruits. When time afforded, the cook would also bake bannock, bread, and cakes with the flour, raisins, sugar, and salt he carried. For the most part, Bridgland worked in areas that were designated parks or reserves in which the wildlife was protected and could not be killed for food. Game birds appear to have been an exception to this rule, for the men would indulge in grouse on the rare occasion when they were able to catch one. A common mountain survival tactic among them was the trapping of an aptly-named fool hen, with a shoelace tied in a slip knot at the end of a stick. Wild raspberries, service berries, blueberries, and huckleberries also occasionally found their way into the cook's menus, baked in pies or served on their own. Canned corned beef did not turn up very often and, amidst the natural fare, probably would not have been appreciated had it done so with any regularity.

A second highlight for the surveyors that summer was the honour of receiving the visit of Sir James Outram (1864–1924), who had made a name for himself the previous year by achieving the first ascent of the Canadian Matterhorn, Mount Assiniboine (11,871 ft [3,618 m]), on the border of Banff Park and British Columbia's Mount Assiniboine Provincial Park. Outram agreed to join the survey party on its trip to the Incompleux River by way of the trail along Flat Creek. On 20 September, Outram, the surveyors, and son Oliver Wheeler, plus five pack ponies and their supplies, were loaded onto a box car that took them twenty-three km up the tracks to Flat Creek, where they were dropped off around midnight. In the morning, they pushed south to the head of the creek. There they established a number of photographic stations, including one on a peak Wheeler named Mount Oliver after his son who "was probably the first white man to set foot on the peak, certainly the first of that size."[28]

On 22 October, with western winds driving snow over the mountains, Wheeler declared the season over. Bridgland returned to Calgary on the train. In his annual report to the Department of the Interior, Wheeler noted that 1902 had been a bad year for surveys, owing to persistent rain, flooded rivers, and foggy, wet conditions. Despite the weather, the party had managed to summit nineteen mountains, nine of which had never before been ascended. They had also climbed several mountains only part-way up, crossed many glaciers, forded innumerable

DESPITE THE WEATHER, THE PARTY HAD MANAGED TO SUMMIT NINETEEN MOUNTAINS, NINE OF WHICH HAD NEVER BEFORE BEEN ASCENDED.

rivers and streams, and spent weeks on end hiking through dense forest, frustrating brûlé (burnt-over, downed forest), and not enough pleasant grassy alpine meadows.

Aside from his mountaineering and packing, Bridgland had readily learned to channel his initiative when bad weather prevented climbing. He would develop plates, write up reports and notes, or do whatever chores he spotted needing to be done. Edward Oliver Wheeler, who many years later became surveyor general of India and undertook phototopographic surveying on Mt Everest when he accompanied Mallory's expedition in 1921, remembers that Bridgland "was tireless in the every day routine of survey life, the pitching of camp, packing of horses, cutting of wood and so on." He further remembers Bridgland's exceptional patience: in the 'office' the sometimes most exasperating chore of changing photographic plates

BRIDGLAND "WAS TIRELESS IN THE EVERY DAY ROUTINE OF SURVEY LIFE, THE PITCHING OF CAMP, PACKING OF HORSES, CUTTING OF WOOD AND SO ON."

> in the dark entirely by feel (with a stuffy dark tent thrown over the head and body) left him unperturbed and like all his work was done with great care and thoroughness. I can remember as a youngster dropping off to sleep with "Bridge" still struggling with plate changing, or writing up the day's notes by candle light. He taught me much in patience, as in climbing and surveying.[29]

✳ *Surveying the Main Range of the Canadian Rockies* ✳

BRIDGLAND SPENT SEVEN SEASONS as Wheeler's "Assistant," although this title hardly represents the work he did, leading a small party of his own for most of each summer.[30] Gradually, his own technique emerged. This reflected a more fastidious approach to the work, for he regularly shot from far more stations than Wheeler would have over the same terrain. From 1903 to 1907, he surveyed the Main Range of the Rocky Mountains with Wheeler and his brother Hector, and worked in the DLS offices in Calgary and Ottawa, compiling survey data and preparing maps in winter. In 1903, they surveyed mostly along the western border of Banff Park, and accomplished first ascents of Mt Sheol (9,108 ft [2,776 m]), Boom Mtn (9,056 ft [2,760 m]), and Mt Bosworth (9,092 ft [2,771 m]). The next summer took the survey through parts of Yoho Park, the northern end of what seventeen years later would become Kootenay Park, and, again, the Sawback Range on the west side of Banff Park. During that summer, Bridgland hiked through two of the world's most famous fossil sites, the Mt Stephen beds and the Burgess Shale site.

In 1886, railway workers had brought the Mt Stephen beds to the attention of veteran Geological Survey of Canada geologist and surveyor George McConnell (1857–1942) and Otto Julius Klotz (1852–1923), a co-founder of the astronomical branch of the Department of the Interior. Klotz made a small collection and sent it to a professor of geology in Michigan who published a paper about discoveries of new trilobites in the Canadian Cordillera.[31] The attention these fossils garnered led the famous Canadian geologist George Mercer Dawson (1849–1901) to take an interest in the area. His interest inspired Charles Doolittle Walcott (1850–1927), secretary of the Smithsonian Institution in Washington, who visited the Mt Stephen beds in 1907. Two years later, Walcott was travelling on horseback between Wapta Mtn and Mt Field when his way was impeded by a large block of shale that had tumbled onto the trail. Reaching for his rock hammer, Walcott split the slab open and discovered the fossils that led to the declaration in 1981 by the United Nations Educational, Scientific, and Cultural Organization (UNESCO) of the Burgess Shale as a World Heritage Site. "Most fossil remains result from the imprint of a hard structure (a shell or a skeleton) being preserved. This site, with its invertebrate fossil remains is unique."[32]

Walcott spent the next twenty years travelling through the Rockies, and in 1922 Bridgland had the opportunity to chat with him about his discovery when Walcott dropped into the Firlands survey camp in Kootenay Park. By then in his seventies, Walcott was still travelling with a pack outfit. He told of how he had followed the fossils' trail from the Colorado canyon all the way up to the Pipestone Pass near Lake Louise. Bridgland, whose keen interest in geology was always advancing, must have been thrilled to meet Walcott and to share stories with him of adventures in the mountains.

In the spring of 1905, Bridgland wrote and passed his final DLS examinations, and on 10 March the Department of the Interior issued him with a standard measure, with which he was to verify the length of his survey chains. This symbolic act conferred on him the office of Dominion Land Surveyor and demonstrated the department's faith that the calculations he submitted were accurate and not in need of correction by the inspector of surveys. Despite his new official qualifications, he still worked with Wheeler's party, through the main

valleys and up the principal peaks of Yoho Park.
The area is one of exceptional beauty that
possesses the variety of mountain peaks,
ridges, and passes that appeal to mountaineers. It was duly noted as a possible
location for an annual camp of the
alpine club that Wheeler had plans to
form.

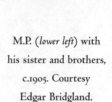

In 1906 and 1907, Bridgland worked in
the Main Range either along or west of the
border between Alberta and British Columbia.
The 1907 work took Morris to Beavermouth, British
Columbia, north of Golden, and up the Blaeberry River, about as old
a known feature as exists on the Pacific Slope in Canadian territory, for
David Thompson and his family used the valley when crossing the
Rockies a century earlier. Early Canada's pre-eminent surveyor had
travelled across Howse Pass and down the valley of the Blaeberry (his
name for it was Portage River) in June 1807. M.P., gradually building
for himself a career that would merit the name Mapper of Mountains,
was toiling in a grand tradition of Canadian surveyors that summer![33]

M.P. (*lower left*) with
his sister and brothers,
c.1905. Courtesy
Edgar Bridgland.

✳ *Marriage and Family* ✳

ON ONE OF BRIDGLAND'S WINTER TRIPS EAST, he met Mary
Elizabeth Perkins, the woman who was to become his wife on 8 January
1908. Like him she was born in 1878. Her father, Charles Perkins,
owned a grocery store north of Toronto in Barrie, Ontario. Mary lived
there until she married and moved to Alberta. Her cousin, Ina Perkins,
had already moved from Toronto to Calgary with her mother and her
brother George, so her move west did not take her away from all
family.[34] Mary moved west but not for the mountains. She much
preferred urban comforts, and so annually she seldom saw her husband
between June and October. Not athletically inclined, she ventured into
the mountains for an extended time only twice, to attend the Alpine
Club of Canada annual camp. Even then, she insisted on sleeping at
the nearest hotel or inn rather than bivouac in the brush-filled tents
used by most members. She bore two sons, Charles and Edgar, in 1908
and 1917, respectively, so she was also occupied with the care of them
through most of M.P.'s years in the mountains. In her younger son's

words, Mary was "a truly Victorian woman"; her priority lay with her family and domestic affairs.

Mary was socially inclined but her rather shy husband was not, so the couple's chief society consisted of visits with their only relatives in the West. Mary's cousin Ina and Frederick Hess had one daughter, Margaret, of whom the Bridglands were fond. She was born in 1916, almost two years before Edgar, and the two were playmates when very young. At the age of eighty-five Margaret Hess fondly recalled the man she called "Uncle Morris," noting that she was treated by him more as a daughter than as a cousin. She remembers especially her uncle's animated lectures on mountain history, geology, botany, glacial activity, even cultural history. Although these topics did not attract or hold interest for her at the age of sixteen, she very much enjoyed the nights her uncle would host lantern slide shows of photographs he had taken on surveys and various vacations.[35] With images of wildlife, beautiful landscapes, striking alpine flowers, various personages, and landscapes in the Rockies, these shows would often be accompanied by impromptu lectures on flora and fauna, geology, or human history.

Margaret also remembers him fondly as her math tutor while she attended St Hilda's School for Girls (1904–1949), Calgary's first non-denominational independent school, located on the 8th Street, 12th Avenue block. She claims that Uncle Morris was responsible for her gaining entrance to university. Doubtless, Bridgland would have been proud had he lived to see that she went on to great achievement, including the honour of being made a member of the Order of Canada.[36] Meanwhile, her father, Frederick, had interests in the lumber industry, and the two men would play billiards together and discuss timber resources in the areas Bridgland had surveyed, in particular the Brazeau River area of Alberta.[37]

Winters were times of physical relaxation although Bridgland was far from idle. He would spend most days at the Calgary office working on triangulations and other map work, and on his days off he would put his personal negatives and photos in condition. In the years when he travelled east, he would take his family with him in their Franklin, a big open car, or one of the two other automobiles that the family later owned. It was normal for him to pitch a tent every night but difficult and worrisome for his wife, who was not used to sleeping on the ground, let alone cooking over a camp stove, washing dishes and clothing in lakes and rivers, and fretting over her children's catching cold or being bug-bitten while they slept. When Bridgland was tied up with work in Calgary, Mary preferred travelling with her children by train to Thunder Bay, then called Fort William, and Port Arthur, where they would embark on a steamer as big as a small ocean liner, which sailed them for thirty-six hours across lakes Superior and Huron to Port McNicoll on Georgian Bay. From there, they would motor across the height of land to Lake Simcoe and her parents' home in Barrie.

✳ *The Alpine Club of Canada Is Born* ✳

REGARDLESS OF THE POTENTIAL DANGERS OF MOUN-TAINEERING, M.P. was always enthralled by the sport that his work required. Clearly, he suffered a pretty bad bite from the mountain bug during the years of his career with the DLS. After only one summer of climbing, he took out a membership with the American Alpine Club and was actively involved in the formation of the Alpine Club of Canada (ACC) three years later, in 1906. He acted as a head guide at the club's first summer camp that year, served in the position of chief mountaineer, documented in the annual volume of the club's *Canadian Alpine Journal* the outings and successes of those who passed their test climbs, and remained an active member of the organization until he died.

At the turn of the twentieth century, most Canadians had only a slight knowledge of the geography of the Canadian portion of the Cordillera, that spine of mountains beginning at the Arctic Ocean and continuing right down through South America. Elizabeth Parker (1856–1944), the founding secretary of the Alpine Club of Canada, wrote in the first issue of the *Canadian Alpine Journal* that those whose primary research and recreational interests lay in the mountains "had long felt the reproach of Canadian apathy to Canadian mountains."[38] While English and American tourists began visiting the Canadian mountains as soon as railway lines gave them access, few Canadians explored or climbed in the immense alpine regions of Alberta and British Columbia. Almost the only exceptions were the geologists, botanists, surveyors, and others whose work it was to document the mountains' features and resources. Although many were members of foreign clubs such as the Appalachian Mountain Club, the American Alpine Club, or the Alpine Club (England), they lacked a formal Canadian mountaineering organization. Beginning in 1906 the Alpine Club of Canada would answer that lack.

WHILE ENGLISH AND AMERICAN TOURISTS BEGAN VISITING THE CANADIAN MOUNTAINS AS SOON AS RAILWAY LINES GAVE THEM ACCESS, FEW CANADIANS EXPLORED OR CLIMBED IN THE IMMENSE ALPINE REGIONS OF ALBERTA AND BRITISH COLUMBIA.

Group of Founders of the Alpine Club of Canada, Winnipeg, 1906.
Left to right back row: unknown; Jack Otto, outfitter; A.O. Wheeler, president; Tom Wilson, member of advisory board;
S.H. Mitchell, assistant secretary; R. Campbell, outfitter. *Left to right front row:* Dan Campbell, outfitter;
M.P. Bridgland, surveyor and chief mountaineer; unknown; Rev. J.C. Herdman, vice-president (Calgary);
A.P. Coleman, vice-president (Toronto). Courtesy Glenbow Archives NA–673–20.

In November 1905, a group of engineers, scientists, and recreational alpinists pondered how they might promote mountaineering as a national sport. As news of the desire to form a Canadian club circulated, the promoters soon learned that mountaineering appealed to a considerable number of Canadians, but they simply had had neither the opportunities and means nor the know-how to climb. Hundreds wrote to express their interest and to enquire about attending the first annual camp. At the inaugural meeting at the Winnipeg YMCA 27 March

1906, M.P. Bridgland and the assembled men and women agreed on the following constitutional objectives for their organization:

(a) The promotion of scientific study and the exploration of Canadian alpine and glacial regions.

(b) The cultivation of Art in relation to mountain scenery.

(c) The education of Canadians to an appreciation of their mountain heritage.

(d) The encouragement of the mountain craft and the opening of new regions as a national playground.

(e) The preservation of the natural beauties of the mountain places and of the fauna and flora in their habitat.

(f) The interchange of ideas with other Alpine organizations.[39]

Although the club is best known as a mountaineering organization, climbing mountains did not represent all members' chief interest. So the club established five grades of membership: honorary members, elected for their service or distinction in mountaineering, research, or exploration; active members, who achieved at least one ascent of at least 10,000 ft (3,048 m) above sea-level; associate members, who desired affiliation with the club but who had not participated on a graduating climb; graduating members, who had not yet qualified as active members but desired to do so; and subscribing members, who did not participate in the outdoor activities of the club but wished to receive its publications. As the ACC meant to be an *alpine* rather than a *mountaineering* club, a climb that qualified one for active membership had to be defined: the route up to or beyond 10,000 ft had to include snow and ice work on a glacier. In order to make such a qualifying climb more appealing to novices, the ACC would opt for routes with rest stops at intermediate elevations where climbers could halt if they decided not to continue to the summit. However, the club also introduced an incentive: those who reached an altitude of at least 9,000 ft would pay reduced camp fees—one dollar per day rather than two.

✳ *The ACC's First Annual Summer Camp, 1906* ✳

THE DAY FOLLOWING THE INAUGURAL MEETING, twenty-seven-year-old M.P. was elected to the committee responsible for organizing the first annual camp. Aside from needing a place near a peak of over 10,000 ft, the camp would require a meadow or clearing large enough to accommodate the forty large platform tents and the 200-ft-long open dining tent that the committee envisaged needing. Yoho Valley was settled on for the base camp and the mountain named Vice President (10,060 ft [3,066 m]) for the official graduating climb (Matterhornist Edward Whymper [1840–1911] had named the mountain in 1904 for David McNicoll, vice-president of the CPR). Over the previous two years, Bridgland and Wheeler had surveyed the area from various camera stations on some of the mountain peaks to the south, and they knew that in addition to its aesthetic appeal it would be relatively easy to access from the railway.

In the first few years of the ACC, President Arthur Wheeler, his brother Hector, and M.P. were granted leave by the Department of the Interior to establish, manage, and take down the annual camps. Bridgland thus occasionally found himself uncomfortably circumstanced, serving as an ACC official but under the eye of the boss of his work. In later years, when sharing with his survey assistants stories about the early days of the club, he often spoke of Wheeler's ironic remarks about how the day ahead promised to be an easy, relaxing one. If it proved to be for others, invariably it was a difficult one for M.P. Although Wheeler has historically been regarded as the most prominent pioneer of the club and its annual climbs, his role was more administrative than Bridgland's. The considerable physical work required in preparing for the accommodation of the guests, then guiding them up graduating climbs, and finally dismantling the camp all fell to M.P., Hector, and anyone whom they could recruit.

By chance, in the year of the inaugural annual camp, M.P.'s classmate Charlie Sissons had taken a teaching position in Revelstoke and was looking for something to do during his summer holiday. Bridgland knew from playing Association Football with Sissons that he was strong and athletic, so he recommended him to A.O. as an able and willing body. Wheeler invited Sissons to join the survey crew as it set up the ACC camp. Excited at the prospect of a mountaineering vacation, Sissons suited himself out in the latest chic mountaineering equipment: a pair of Fox's spiral puttees, shoes nailed in the Tyrolean fashion, a pair of Hudson's Bay Company blankets, and a Swiss ice-axe on which he had his name and the date inscribed. However, only after days of backbreaking work erecting and brushing tents in preparation for the camp did Sissons finally climb his first peak and graduate as an active member of the club.

EXCITED AT THE PROSPECT OF A MOUNTAINEERING VACATION, SISSONS SUITED HIMSELF OUT IN THE LATEST CHIC MOUNTAINEERING EQUIPMENT: A PAIR OF FOX'S SPIRAL PUTTEES, SHOES NAILED IN THE TYROLEAN FASHION, A PAIR OF HUDSON'S BAY COMPANY BLANKETS, AND A SWISS ICE-AXE ON WHICH HE HAD HIS NAME AND THE DATE INSCRIBED.

M.P., Charlie, and Hector's preparation for the camp included lumber-jacking and axe-work. They built an enormous fire pit and log benches to surround it, cut poles for tent supports, set up the huge canvas awnings of the enormous dining tent and some forty sleeping tents, made tables and benches for the dining area, and cut and gathered soft boughs with which to "brush" the tents, that is, make sleeping pads for the guests from the springy branches of lodgepole pine or other evergreens. "Bridgland was a tower of strength in this work. ...[H]e never lost sight of the object for which the survey party was lent—to make members as comfortable as possible and to graduate as many as possible as safely as possible."[40]

M.P. and his colleagues started at the beginning of July to prepare the camp; nine days later, guests began disembarking from the train at Field, British Columbia. From there a crude corduroy road took them through the forest for about eleven km to the shores of Emerald Lake. Most members gladly hiked in, and the few who followed behind in horse-drawn wagons had to abandon these at the lake, for from there the trail to the camp continued over a broad glacial delta, then up a steep cliff wall before at last reaching the summit of Yoho Pass. From the height of the pass, the sight of a neat "city" of white tents clustered in the valley greeted the incoming campers. Tent City was set out in organized fashion: the sleeping tents and living quarters of the men and women, respectively, were dubbed Residence Park; Official Square included the dining, kitchen, and tea tents as well as the official flagpole and fire pit; and the Paddock held the horses. In Official Square, a bulletin board informed the guests of daily activities and advertised the climbs on which places remained available. Although the Boer War was long over and the nightmare of the First World War yet to begin, the military discipline about the camp was unmistakable.

As the club's founding chief mountaineer, Bridgland officially joined the ranks of the Canadian and European mountaineering elite. Although co-workers knew about his climbing prowess, this office with the ACC expanded his reputation to a much wider group. For instance, he guided alongside such distinguished alpinists as Edward Feuz Jr and Gottfried Feuz.[41] Two days prior to the first official climb, M.P., the Swiss guides, a Rev Dr J.C. Herdman, and P.D. McTavish set out to climb Vice President looking out for potential problems that their less experienced charges might encounter. With their combined expertise,

AS THE CLUB'S FOUNDING CHIEF MOUNTAINEER, BRIDGLAND OFFICIALLY JOINED THE RANKS OF THE CANADIAN AND EUROPEAN MOUNTAINEERING ELITE.

Alpine Club of Canada's First Summer Camp, 1906. M.P. is in the centre of the front row, resting on his left elbow. To *his* immediate left, resting on their right elbows, are Gottfried and Edward Feuz. To his right on the far side of the climbing rope, also resting on his right elbow, is Hector Wheeler. The woman in white blouse and hat behind Bridgland is Clara Wheeler, first wife of A.O. Wheeler, who stands behind her and to her left, wearing a toque and adorned by his pipe. The first woman to *his* left is Elizabeth Parker, founding secretary of the ACC.
Courtesy Whyte Museum of the Canadian Rockies V200/ PA–44–13, Edward Feuz fonds.

they had no trouble choosing a varied but safe route. At the camp's bulletin board on Tuesday 10 July, M.P. met the first group of graduating members, seven men and two women. Under his lead, they set out, making the return trip to the summit in under ten hours, all of them rainy.[42]

In the first camp, members were enthralled by the institution of the evening campfire, which encouraged and fostered congeniality and patriotism among participants, and which became a club tradition.

Guides in Charge of Climbing—Yoho Camp 1906. *Left to right:* Edouard Feuz, Swiss guide; H.G. Wheeler, asst. guide; M.P. Bridgland, chief guide; Gottfried Feuz, Swiss guide. *Canadian Alpine Journal* 1 (1907/08), facing 168.

EVEN THOSE WHO DID
NOT CLIMB BUT CAME
TO THE CAMPS IN
SUPPORT OF THEIR
SPOUSES OR FRIENDS
ATTENDED THE
EVENING CAMP FIRE,
SANG ALONG, SAT AND
LISTENED TO STORIES
FROM THE VETERAN
ALPINISTS IN
ATTENDANCE, AND
EXCHANGED STORIES
OF THEIR OWN.

Even those who did not climb but came to the camps in support of their spouses or friends attended the evening camp fire, sang along, sat and listened to stories from the veteran alpinists in attendance, and exchanged stories of their own. One contributor to the club's *Canadian Alpine Journal* confessed that after hearing so many of Jack Otto's bear stories the mere sight of a porcupine made him believe he was face to face with a grizzly.

Over the course of the camp, fifteen women and twenty-seven men graduated to active membership. Of these, twenty-nine members qualified on Bridgland's rope. Although M.P. had to repeat the same climb day after day, Sissons doubted that "any one of the scores of members who qualified on his rope would have noticed any sign of impatience at what must have proven at times a tedious experience."[43] As well, because M.P. oversaw his friend's apprenticeship, Sissons was soon able to lead parties of his own:

Bridgland was a man of few words, and I do not recall that he took me aside to instruct me in the art, or that on many occasions he discussed routes of attack with me. I simply learned as I followed him, and, after those first two months with him in 1906, I myself was usually in charge of a rope when I climbed with the Alpine Club. In our survey work (indeed in our Alpine Club ascents at that date) we had not recourse to such modern gadgets as pitons and crampons and rope-ladders. We were content with our Swiss ice-axes, our stout boots, solidly nailed around soles and heels in Bridgland's case and more lightly Tyrolean fashion in my case, and a climbing rope.[44]

Climb of Mt Vice-President 1906,
Canadian Alpine Journal 1 (1907/08), facing 171.

Aside from the official climbs that year, ascents were made of Mt Burgess (8,527 ft [2,599 m]), Mt Wapta (9,115 ft [2,778 m]), Mt Collie (10,224 ft [3,116 m]), Mt Field (8,672 ft [2,643 m]), Mt Marpole (9,833 ft [2,997 m]), and a remarkable route up Marpole known as Amgadaom Point, an acronym of those whose first ascent of Marpole put up this route, A. M. Gordon, A. Dunn, and A.O. McCrae. Chief Mountaineer Bridgland's responsibilities included writing a report of all the climbs attempted, lists of the participants, and the duration of each ascent. The "Report of the Chief Mountaineer" was published in the annual volume of the *Canadian Alpine Journal*, along with other official reports, camp stories, and scientific articles about Rocky Mountain botany, geology, and glaciology.

✳ ACC Summer Camps, 1907–1911 ✳

THE FIRST YEAR'S EFFORT, Herculean as it must have been for Bridgland, quickly became the norm rather than the exception. The 1907 camp was located at Paradise Valley, south of Lake Louise in the Bow Range. Again, Bridgland was instrumental in the set-up, guiding, and take-down. Because spring arrived particularly late, the planned site for the base camp lay under a cover of snow when the time came to establish it. As a result, Morris and a gang of workers loaned by the CPR felled trees in the lower valley at the base of Mt Aberdeen to make a clearing. (Evidently, what was permitted in a national park in the early days differed widely from acceptable and indeed legal practices today.) These trees were then sized for tent poles, dining tables, benches, and firewood. Two peaks, Mt Temple (11,625 ft [3,543 m]) and Mt Aberdeen (10,342 ft [3,152 m]), were chosen for gradu-

Corner of the Alpine Club of Canada's first Club House, Banff. From A.O. Wheeler, *The Selkirk Mountains: A Guide for Mountain Climbers and Pilgrims* (1912), facing 164.

ating climbs. The bulletin in Official Square posted the 5:30 am and 6:30 am departure times for the respective ascents; most of the candidates chose the lower Mt Aberdeen. As he had the previous year, M.P. served as chief mountaineer and guided many of the sixty-six who graduated to active membership at the 1907 camp. Despite the bad weather, it attracted 150 members and guests; not one member who attempted to graduate failed to do so. Sissons took charge of a second camp at Lake O'Hara, and Bridgland welcomed another good friend, L.C. Wilson, into the fold of the club. As a bookstore owner from Calgary, Wilson had had few opportunities to climb, but under Bridgland's guidance he safely graduated to active membership and returned year after year. The nickname "Bridge" that Wheeler gave him because of his notable strength and endurance also proved apt for his involvement in the ACC: he so often helped people over difficult terrain, offering his knowledge, his patience, and his steady physical support to any of the club members who wanted to attempt a climb. At the 1907 camp, as well, the annual meeting held around the evening campfire resolved to build a club house at Banff, on Mountain Avenue part-way up the road to the Upper Hot Springs.[45] A second point of order was to award to Wheeler and Bridgland each a pair of men's and ladies' ice-axes in recognition of their winning photographs in the club's annual contest.

HE SO OFTEN HELPED PEOPLE OVER DIFFICULT TERRAIN, OFFERING HIS KNOWLEDGE, HIS PATIENCE, AND HIS STEADY PHYSICAL SUPPORT TO ANY OF THE CLUB MEMBERS WHO WANTED TO ATTEMPT A CLIMB.

After one of the guides' nightly debriefings of the day's activities that summer, a tired M.P. crept to his tent and lit a match to avoid tripping over any of the inmates. In the dim light he found not only his bedroll spread out on his mattress of boughs but also a large porcupine happily nestling on his pillow. Tired as he was, and without any other place to lay his head, he prodded the animal gently with an ice axe. Not too concerned with having to move, the porcupine got up and settled himself on a nearby head, which, luckily, belonged to a sufficiently sound sleeper that he did not awaken. With an avid tooth for anything salty, porcupines numbered mountaineering boots among their favourite snacks. As a result of the danger of having one's boots gnawed to bits if left out, these were often stowed under their owner's pillow; perhaps this arrangement prompted the creature in question to land up where it did.

Rogers Pass in the Selkirks was chosen as the site for the ACC's 1908 camp. Because this area receives much more precipitation than the Rockies farther east, the permanent snow line descends to much lower elevations, and the vegetation is generally richer, so the peaks in this area have softer outlines than the stark rocky promontories that dominated the horizon at the previous two years' camps. The graduating peaks chosen for the year were mounts Rogers (10,551 ft [3,216 m]) and Hermit (10,151 ft [3,094 m]). Yet again, and even though he had married earlier in the year, M.P. served as chief mountaineer, guiding most of the fifty-seven graduating members up their first peak of more than 10,000 ft. Unlike the day trips of previous years, the climbs in 1908 took two days; the hut at the tree line on Mt Rogers afforded overnight shelter. Besides the anticipated duties, M.P. also had, as chief mountaineer, to retrieve the body of Miss Helen Hatch, the club's first fatality. She had been climbing with Wheeler's party on Avalanche Mtn (9,386 ft [2,861 m]), a lower, easier peak south of camp. Without giving her guide time to stop her, she ran past him and slipped over a precipice, falling to her death. This event greatly sobered the participants.

In time the club's business returned to normal. When it did, M.P. was elected at the annual meeting to a three-year-term as a co-vice-president, along with J.D. Patterson. Although he took on increasing responsibilities as a surveyor in the 1909 season, M.P. remained the chief mountaineer of the ACC's annual camp, held at Lake O'Hara, in Yoho Park. As the reputation of the club grew, so did its membership: that year, climbers arrived from Scotland, England, the United States, and the Netherlands. The club was also honoured to receive Conrad Kain (1883–1934), the famous Austrian alpinist who would later write up his adventures in the Canadian Rockies in the popular book *Where the Clouds Can Go* (1935), and who assisted with some of the guiding.[46] Having learned their lessons about the joys of camping in snow the previous summer, M.P. organized the camp in August rather than July. Again, two peaks were chosen for graduating climbs, but before the guests arrived, when Bridgland and the other guides set out to determine the routes they would use with the graduating members, they found that Mt Odaray (10,365 ft [3,159 m]) had two chimneys that would present great difficulties to novices, so they settled on Mt Huber (11,050 ft [3,368 m]) as the sole peak for graduating candidates that season.

When the summer camp of 1910 was held at the Consolation Lakes (southeast of Lake Louise), a change was forced upon it, one that precluded the management of it by either Wheeler or Bridgland. For the first time, the Department of the Interior refused to allow the men leave from their professional duties. This denial so infuriated Wheeler that, although he managed a year's exemption, he quit the service of the federal government the following year when permission was again denied. Thereafter, he worked only for the provincial governments of British Columbia and Alberta; both allowed him the leave he needed to retain his stature in the ACC. Bridgland, who remained with the DLS, did manage to return to the ACC camp in 1911, when he took his wife and Charles, the first of their two sons, to the camp at Sherbrooke Lake, just west of the border between Banff and Yoho parks and east of the Yoho River valley. Despite the beautiful setting, Mary remained unconvinced that the mountains were worth climbing, and she contented herself with socializing in the tea tent during the day and around the campfire at night, while her husband assumed his regular prominent role.

AS THE REPUTATION OF THE CLUB GREW, SO DID ITS MEMBERSHIP: THAT YEAR, CLIMBERS ARRIVED FROM SCOTLAND, ENGLAND, THE UNITED STATES, AND THE NETHERLANDS.

✳ *Bridgland the Consummate Guide* ✳

FOR HIS PART, M.P. must have had good insight into the workings of administrations: rather than confront the Department of the Interior as the fiery Wheeler had done, he arranged to attend several subsequent ACC camps on his own time or when a camp happened to occur in proximity to his assigned summer's survey work. Still, his involvement was constrained: after 1910 he no longer set up camps or served as chief mountaineer. Meanwhile, during winters he would give illustrated lectures on various trips or areas of interest at ACC sectional meetings in Calgary.

Whenever visitors, especially easterners, came to visit the Bridglands in Calgary, M.P. would take them to Banff and on day hikes from there or, if they were so inclined, guide his guests to the peak of a nearby mountain. Often he would bring family and friends to the ACC camps so that if they were not interested in climbing they would have a good time socializing. In 1929, he somehow convinced Mary to attend another annual camp, at Rogers Pass. Her cousin Ina Hess was not of the same ilk, however; when she came out to the camp with her husband and daughter, she also completed a graduating climb and became an active member of the club, much to M.P.'s gratification. He would contribute painstakingly and generously to the education of novice mountaineers, and so he adopted a different technique from the trial-by-fire one used to introduce surveyors to mountain climbing. As he did with ACC members, he would teach his friends all the practical knowledge they needed, including the proper handling of ropes and even the way in which to use an ice axe as a rudder when glissading. Clearly, he wanted to make the sport accessible to everyone.

Of chosen destinations in the mountains and foothills, one of M.P.'s personal low-level favourites was the site of the Old Peigan Post, or Old Bow Fort. Margaret remembers many trips back to this place, which was not marked on any map but which he could locate by finding the broken tree with a particular bird's nest by the side of the road on the way through Morley, Alberta, to Banff Park. From the tree, one had to trek into the forest a little distance to a tree bearing a surveyor's blaze dated 1874, which was later cut down "by some scallywag." The post was originally a Hudson's Bay Company fort, established in 1832 on the Bow River near the confluence of Old Fort Creek, on the Morley Reserve west of Morley itself and east of Seebe.

HE WOULD CONTRIBUTE PAINSTAKINGLY AND GENEROUSLY TO THE EDUCATION OF NOVICE MOUNTAINEERS, AND SO HE ADOPTED A DIFFERENT TECHNIQUE FROM THE TRIAL-BY-FIRE ONE USED TO INTRODUCE SURVEYORS TO MOUNTAIN CLIMBING.

The post was abandoned when, around 1835, it burned to the ground. When Bridgland visited it on a short half-day trip from Calgary, its crumbling and blackened stone chimneys still stood, and some rotting wooden axles and other remnants remained from the many travellers and settlers who had camped at the ruins on their way to other destinations.

Charles and Edgar Bridgland and Margaret Hess were often brought out on field trips, although the latter two were not avid rock climbers. Edgar did not mind climbing up mountains, but coming down them was another matter. He found it quite a challenge to keep up to his agile father and long-legged older brother, eight years his senior. Margaret was a keen hiker but "Uncle Morris" would not permit a young lady to hike in trousers, so she had to content herself with activities that she could accomplish in a skirt or dress. Sometimes she would help her uncle in his nature studies by holding flowers he was photographing to give an idea of scale, or she would stand in a particular place so that she cast a shadow for him in order to throw the requisite shade on his subject. Bridgland's attitude towards women of whom he was protective evidently was either unenlightened or marked by the death of Helen Hatch or both.

Despite elitist claims among some cosmopolitan alpinists that to popularize the sport was to vulgarize it, ACC members maintained that "it is the people's right to have primitive access to the remote places of safest retreat from the fever and the fret of the market place and the beaten tracts of life."[47] Understanding of most people's desire to climb mountains, Morris was able to look after groups of climbers with different expectations and levels of expertise, and make the experience accessible and enjoyable. Edward Wheeler confirmed Bridgland's aptitude as a guide:

> He led party after party up the same climb, to him an easy and uninteresting climb, always safely, always with patience and good temper, and always with the utmost consideration for his party. On an occasional "rest" day he would gratify his own enthusiasm by undertaking difficult climbs or new routes with the Swiss guides or with some experienced companion. I have met few men who so consistently subordinated their own inclinations to the achievement of the object.[48]

UNDERSTANDING OF MOST PEOPLE'S DESIRE TO CLIMB MOUNTAINS, MORRIS WAS ABLE TO LOOK AFTER GROUPS OF CLIMBERS WITH DIFFERENT EXPECTATIONS AND LEVELS OF EXPERTISE, AND MAKE THE EXPERIENCE ACCESSIBLE AND ENJOYABLE.

He was recognized by his mountaineering peers as having few equals, particularly on rock work, and their good opinion issued in part from his resolute refusal to sacrifice safety in order to satisfy curiosity. Whether routes were difficult or easy he stuck to one policy in choosing them: "Better the devil you know, than the devil you don't know."[49] Those who followed him as their guide knew that they could depend on him. Moreover, guiding uncovered some aspects of his personality that surveying did not. Indeed, the good reputation that the ACC enjoyed among alpine clubs from its inception doubtless owes much to the man's disposition. In *The Alpine Herald*, a newspaper published after the first ACC camp, the women campers wrote a column to thank the guides for their generosity and chivalry: "Sometimes they applied to our boots the nails that made more sure our footsteps on these rough heights.... Sometimes these same boots were greased for us, and again the grease was merely lent, and, in response to our fervent thanks, came the answer, 'It's all right, it's not mine, it belongs to Bridgland.'"[50] Teasing is a high compliment when it comes to assessing character. Such generosity and geniality as M.P. evidently routinely exhibited go far towards establishing a positive atmosphere in any club. No source himself of high drama or narrow escapes, he brought a superb temperament to a region of Canada to which others needed only to be guided, so that they could find mountain nature's ineffable beauty and rigorous physical challenges for themselves. Free of ego, he could let the mountains speak for themselves even as he personally guided others through, up, and down them.

FREE OF EGO, HE
COULD LET THE
MOUNTAINS SPEAK FOR
THEMSELVES EVEN AS
HE PERSONALLY
GUIDED OTHERS
THROUGH, UP, AND
DOWN THEM.

✳ *Triangulation Surveys of the Railway Belt, Crowsnest Forest Reserve, and Waterton Lakes Park* ✳

AS WE RETURN TO HIS CAREER WITH DOMINION LANDS SURVEY, we find that in 1908 and 1909, while still with Wheeler, Bridgland surveyed in the Railway Belt at Shuswap Lake and north and south of Golden, British Columbia. In 1909 he took charge of a sub-party that carried out its own daily excursions, and for the first time he was formally recognized and paid as the chief of party. In 1910, he revisited many of the peaks he climbed in his very first season but the work required of him was quite different. He still took photographs from triangulation stations, but now he had to pay special attention to triangulation readings for points along the railway line, and also had to classify the vacant lands in the valleys of the Railway Belt according to their optimal use, whether as orchards, farms, ranches, or logging areas, or what was deemed at the time "nothing whatever."

During the following two years, M.P. held the distinction of primary surveyor for the Railway Belt triangulation surveys, and continued from then on to be considered by the Department of the Interior as its leading authority in Rocky Mountain surveying, particularly in phototopographic techniques.[51] This distinction Wheeler might have garnered for himself had he not resigned from the federal service following the 1910 season. His departure effectively left Bridgland as the most experienced and arguably the most skilled phototopographical surveyor employed by the DLS. Incidentally, Bridgland worked also for the Alberta provincial government as an Alberta Land Surveyor, for which he received his commission in 1911. In reviewing Bridgland's 1910 annual report, Charles Sissons, who became a writer and a prominent educator in Ontario, could not help but notice how the syntax of Bridgland's prose revealed just how much physical energy he was willing to invest in his work, and the good humour with which he did so:

> He disliked rhetoric as much as he disliked all other extravagances. But in looking over some of his reports, I was intrigued by two unusual words in a paragraph, "delight" and "only." The paragraph is reproduced as characteristic of his method:

> On July 3, a start was made for Mt Begbie [8,967 ft (2,733 m)] to the west of Revelstoke. Crossing Columbia River by a bridge at this point, we travelled south about four miles by means of a settler's trail. From here the horses were sent back and we proceeded on foot about three miles farther south to the base of Mt Begbie. Camp was pitched at night on the side of the mountain about 2,000 feet above the Columbia Valley. Much to our delight, the following day was fine and beautifully clear. The mountains offered no difficulty and we were on the summit by nine o'clock. A cairn was erected, five feet in diameter at the base and eight feet, seven inches high. In the rock at the centre of the cairn a hole was built to receive the

brass bolt, and four holes, each distant six feet from the central hole and bearing north, east, south and west respectively, were drilled for reference bolts. This cairn was designated as signal XXXVIII. The trip to the mountain and return occupied only three days.

"The word 'delight,'" Sissons noted,

> is accounted for by the fact that work had been held up by rain, and "only" provoked by the nature of the country and the prodigious effort involved in penetrating the swamp, fallen timber, devil's club, and alder between the end of the road and the base of the mountain. As a matter of fact, the party was back in main camp on noon of the third day. It is significant also of Bridgland's capacity for under-statement that he does not mention a delay at the bergschrund, or the hail storm on top which extended the work there to seven hours.[52]

Work in 1910 occurred in the Selkirks and Monashees, in 1911 in them as well as in the Rockies, and in 1912 back again in the Selkirks and Monashees for triangulation work on the CPR line so that he could connect his base at Salmon Arm with the base that Wheeler had established earlier at Revelstoke. The 1911 season was plagued by bad weather, long stretches of both rain and snow. A mountain suitably dubbed Storm Mtn (10,361 ft [3,158 m]) in 1884 by George Mercer Dawson proved particularly vexing to Bridgland, for he climbed it on three separate occasions only to be prevented from making photographs by devilishly sudden bad weather. From Banff to Revelstoke he chased after good weather all summer long. In fact, his diary for that season makes him sound impatient for one of the infrequent times in his career in the field. Men quit on him, he had to make a quick trip to Calgary to hire others, he wore out a Swiss guide (Christian Hasler Jr or Sr), and he was stuck in a fall of three feet of snow while in a "light" camp high up on Mt Begbie.[53]

In 1913 and 1914, M.P. worked in what was for him a new region— the Crowsnest Forest Reserve, the extensive tract of the eastern slopes of the Rockies that separated Bow River Forest Reserve to the north from Waterton Lakes Park (established as Waterton Lakes Forest Park, 1895) to the south. The northern portion of this vast expanse occupied the

THE 1911 SEASON WAS PLAGUED BY BAD WEATHER, LONG STRETCHES OF BOTH RAIN AND SNOW.

first of these two summers' work. Base camp was established in "The Gap," where the Oldman River flows through the Livingstone Range, about sixty-five km southwest of Claresholm, Alberta. The easiest access was discovered later to run, not from Claresholm, but rather north from Lundbreck, Alberta, on the southerly Crowsnest branch of the CPR. Horses used the previous summer in Golden, British Columbia, and shipped by rail in the spring to Okotoks, Alberta, were brought to the base camp. The party consisted of its chief and Albert Edward Hyatt (1892–1915), his assistant, as well as a cook, packer, assistant packer, and two labourers.[54] The men's wages for 1913 were stipulated as follows: Dominion Land surveyors such as Bridgland received $8 per day, his assistant $3, the cook and packers $2.50, and the labourers or teamsters $1.50.[55]

Work was delayed until 13 June but it continued through to 6 August from this base. After completion of this stage, which took in mountains accessible from Oldman River, Dutch Creek, and Livingstone River and its tributaries, the men moved gradually southward, photographing from the access provided by Racehorse, Vicary, and Daisy creeks, and thereby completing the drainage basin of the Oldman River. The peaks in this

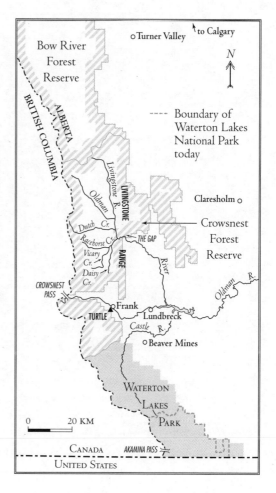

Bridgland's Southern Surveys (with Hyatt), 1913-1914

region's main range measure from 8,500 to 10,000 ft (2,600 to 3,050 m). Summits of the slightly lower Livingstone Range, twenty-five km or so to the east, run only a little less high at some points. Bridgland found that a satisfactory survey that included the railway had to extend beyond the boundary of the Forest Reserve at points. Consistent with his character, he went the extra yard rather than interpreting his orders more narrowly. He even established seven stations south of the railway line in anticipation of connecting up the next season's survey farther south. By 25 September, when the party had made its way back to Calgary and the men were paid off, Bridgland was able to report that

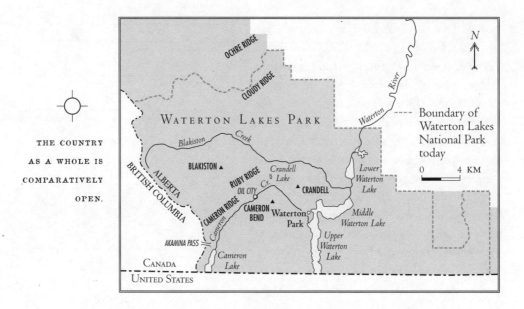

THE COUNTRY

AS A WHOLE IS

COMPARATIVELY

OPEN.

Bridgland's Survey (with Hyatt) of Waterton Lakes Park, 1914

photographic sets had been shot from 114 stations, and over thirty-five km of the railway traversed. In total, more than sixty dozen glass plate negatives were exposed. No smoke whatsoever and only eighteen days of rain had precluded photography. Bridgland's identification of the greatest problem that challenged the men in this region would come as no surprise to anyone familiar with southern Alberta: "incessant high winds, which rendered good work extremely difficult."[56] As well, however, the large number of stations reflects the fact that the region did not lend itself as well as others to phototopographical surveying: "Our worst trouble," he wrote to the surveyor general in the middle of the summer, "is due to a lack of commanding peaks which necessitates more stations than would be needed otherwise."[57]

Further in his annual report to the surveyor general, Bridgland painted an attractive picture of the country. While it did not, as far as he could determine in a single season, abound in game, and although it had been burnt over recently, he noted some of its attractions, not least the famed trout of the upper Oldman River:

The country as a whole is comparatively open. There are good trails which are practically free from steep grades, bad swamps, or muskegs, in all the main valleys. In many cases horses can easily be taken through country where no trail exists. In most sections feed is very plentiful, the valleys as a rule being "U" shaped with large meadows in the bottoms. In addition many of the hillsides are open and are covered with a luxuriant growth of grass and pea vine. During the past summer, the upper part of Livingston valley was used as a range for several hundred cattle. Wild flowers of many varieties are abundant, particularly on some of the upper slopes near timber line. Edible fruits are scarce, soap berries (if these can be called edible) and wild black currants being the only varieties seen in any quantity.[58]

The 1914 season was a busy one for Bridgland; he had, in fact, responsibility for two surveys. The surveyor general wanted him to repeat and complete a triangulation survey in the Railway Belt near Golden that Wheeler had undertaken some years before but left unfinished. Bridgland explains the entire season's work in his report:

My work during the season of 1914 consisted of a topographical survey of the southern part of the Crowsnest Forest reserve and a retracement of the triangulation of the Rocky and Selkirk mountains from Calgary to within a short distance of Golden.

In order to complete these two surveys in one season it was considered advisable to start work on the latter as early as possible. Accordingly on May 12, I engaged one man and commenced this survey which was continued until May 26. … As further work was then impossible, owing to the amount of snow still remaining on the higher peaks of the mountains, I returned to Calgary, engaged more men, and on May 27 proceeded to Lundbreck where the camp equipage and supplies had previously been shipped. From there we proceeded to our first camp…from where the survey of the Crowsnest Forest reserve was carried on until July 6.

Two extra men were then engaged, and the party was divided. My assistant with three men moved south to Beaver Mines and continued the survey of the reserve, while I with four men left for Morley on the main line of the Canadian Pacific railway to continue the triangulation survey.[59]

Photograph by M.P. Bridgland of the view of some of
the remains of the original locaton of Oil City (1901–06)
from Ruby Ridge, Waterton Lakes Park, 1914 (detail).
Courtesy Library and Archives Canada R214.

Photograph by M.P. Bridgland of the view of the
remains of the original location of Oil City (1901–06)
from Ruby Ridge, Waterton Lakes Park, Stn, 124,
no 726, direction southsouthwest, 1914. Courtesy
Library and Archives Canada R214.

Bridgland worked until the end of August
on the triangulation survey in the Bow River
valley and environs, including Spray Lakes
and Lake Minnewanka. Then he and horses
rejoined the crew surveying the Crowsnest
Forest Reserve. Edward Hyatt, the unnamed
assistant to whom Bridgland refers in the
passage above, headed up the second crew. He had worked steadily
south from the vicinity of Crowsnest Pass towards Waterton Lakes
Park in August. All men worked in the Crowsnest area for the month

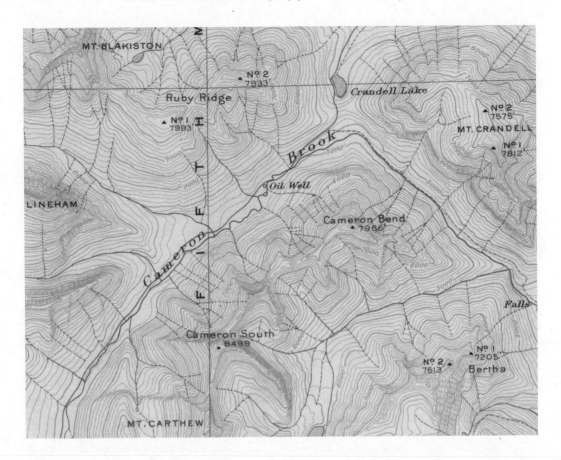

Map of Crowsnest Forest and Waterton Lakes Park, Rocky Mountains Forest Reserve. From Photographic Surveys by M.P. Bridgland and A.E. Hyatt 1913–14. 1:62,500. Ottawa: Dept. of the Interior, [1915] (detail). The "Falls" behind today's town of Waterton appear on the right, and the "Oil Well" on Cameron Brook (now Cameron Creek) is in the centre. Courtesy William C. Wonders Map Collection, University of Alberta.

of September. Bridgland recorded snow and very high wind on Friday 11 and Saturday 12 September, so everyone worked on Sunday, when, despite strong winds, Ochre Ridge and Cloudy Ridge, on the park's northern perimeter, were both occupied. Subsequently, Blakiston Creek was traced and stations named Lakeview 1, 2, and 3 were established in the area of Mt Crandell (7,812 ft [2,381 m]). On Monday 21 September, still with a sense of urgency to get the lowermost edges of the work completed that summer, the two crews occupied Cameron Ridge and Mt Blakiston (9,619 ft [2,932 m]), and a week later had made their way to Ruby Ridge and Cameron Bend, overlooking the site of the first oil well drilled in western Canada. But the season's work stalled when rain came on the first day of October and ended abruptly two days later, when a snowstorm, which "lasted several days, render[ed] further work impossible."[60]

South of the railway in Crowsnest Pass, Bridgland described the country covered by the reserve as somewhat different than north of it. In particular, the main mountain range "loses its distinctive characteristics. It becomes much more broken and several good passes exist, of which the Akamina pass in t[ownshi]p. 1–1–5, is worthy of note, being crossed by a good wagon road." His image of the country looking out from the mountains near Waterton Lakes Park would make it readily recognizable today: "the country to the east becomes much more rugged and broken, peaks from 7,500 to 8,500 feet in height extending to the easterly limit of the reserve, where the change from mountains to very low foot-hills or rolling prairie is very abrupt."

High winds challenged the men again in 1914, but they also faced a summer of worse weather for photography: of the 125 days spent on the survey by Hyatt and or Bridgland, "forty-six were totally lost owing to bad weather, and many other days were partially lost." The reader of his report can almost hear Bridgland moaning, however uncharacteristic this might be. He must have hated doing anything but a first-class job, but the weather conspired against him, especially when he needed its cooperation most: "In September, when trying to complete the work, while only thirteen days were entirely lost, there were but nine fine days during the entire month." Still it was grand country in M.P.'s view: good pasturage and gradual slopes abounded, even if the variety of flowers and grasses encountered in 1913 was unmatched in 1914. He made a point of noting the abundance of wild black currants. He was intrigued as well by the coal mines being worked in the area and by the operation of one crude oil well, once abandoned but then "being worked under new management, and at a depth of 970 feet is yielding fifteen to eighteen barrels of crude oil per day."[61] All in all, the summers of 1913 and 1914 sent the mapper of mountains to territory that exposed him to a different sort of mountainous landscape. The same would happen in 1915. ✳

✳ *Notes* ✳

1 In 1841, *The Act of Union* united Upper Canada and Lower Canada as the Province of Canada. This Province had a legislative union with eighty-four members, divided equally between Canada East and Canada West. These jurisdictions took the names of Québec and Ontario, respectively, at the time of Confederation in 1867.

2 Victoria College gained the right to grant university degrees in 1841, the year when it was established in Cobourg, on Lake Ontario east of Toronto. It was renamed Victoria University College and was federated with and moved to the University of Toronto in 1892.

3 Qtd in George W. Spragge, "Bridgland, James William," *Dictionary of Canadian Biography*, vol. X, ed. Marc La Terreur (Toronto: UTP, 1972), 90–91.

4 A post office marked the location of Fairbank on the 1878 map of North York Township, at the corner of what are now Dufferin Street and Eglinton Avenue, just south of the Parsons' estate on Concession Road III West. Today the name is remembered in Toronto by a small park southwest of the site of the post office.

5 This railway was designed to link lakes Ontario, Simcoe, and Huron, and to facilitate the shipment of timber or mineral resources harvested in Simcoe County. Although it allowed many small businesses to flourish both in Toronto and in the north country, the company soon went nearly bankrupt. It was reorganized a number of times under different names, eventually to be incorporated by the Grand Trunk Railway, and then by the Canadian National system. Nineteenth-century Canada's leading railway surveyor and construction engineer, Sandford Fleming, who lived on a neighbouring farm, and who, as we saw earlier, would make a major contribution to Canadian railway survey history, was the engineer-in-chief of the Northern Railway in 1857.

6 M.P. Bridgland, "Electrical Discharge in Air at Low Pressure," *Acta Victoriana* 24.4 (1901): 361-64.

7 *Acta Victoriana* 21.4 (1898): 213. The quotation is from Alexander Pope's poem, *Epistle to Dr. Arbuthnot* (1735), l. 128. In "numbers," Pope's reference is to the metres of poetry, not to mathematics, so the allusive use of it in this context is either witty or mistaken.

8 In 1906, the school was officially incorporated into the University of Toronto as the Faculty of Applied Science and Engineering.

9 *Calendar of the Ontario School of Practical Sciences, 1902–1903,* 106.

10 *Acta Victoriana* 24.8 (1901): 429. The quotation is a bastardization of "'genius' (which means transcendent capacity of taking trouble, first of all)," a remark by Thomas Carlyle (1795–1881), *The History of Frederich II of Prussia, called Frederick the Great*, 6 vols. (1858–65), vol. iv, chapt. 3, itself an echo with a difference of the maxim coined by Comte de Buffon (1707–88): "La génie n'est qu'une plus grande aptitude à la patience" (Marie-Jean Hérault de Séchelles, *Voyage à Montbar,*

contenant des Details très intéressans sur le Caractère, la Personne et les Écrits de Buffon [Paris: Solvet, an IX (1801)], 15).

11 James Loudon (1841–1916), who went on to become the university's fourth president (1902–06), is characterized as having attained his "principal achievements" "in the promotion rather than in the advancement of science" (James Grant Greenlee, "Loudon, James," *Dictionary of Canadian Biography*, vol. XIV, gen. eds. Ramsay Cook and Jean Hamelin [Toronto: UTP, 1998], 665), so he might seem more likely to encourage young Bridgland than his nephew, William James Loudon, who was also teaching in the SPS in 1901–02. However, Edgar Bridgland, M.P.'s son, has stated that he himself was taught at the University of Toronto by Thomas Richardson Loudon, the son of William James.

It was almost certainly William James Loudon whom Bridgland intended to honour when he nominated Loudon as the name for a creek and mountain in Jan. 1826. At an elevation of 10,568 ft (3,221 m), Mount Loudon stands in the Front Range, Murchison Group, of the Siffleur Wilderness Area, east of Banff National Park, thirty-three km south of the widening of the North Saskatchewan River known as Abraham Lake. Mt Loudon is visible from Hwy 11, also known as David Thompson Highway, running east from the Icefields Parkway at Saskatchewan Crossing to Nordegg, Rocky Mountain House, Sylvan Lake, and Red Deer, Alberta. Loudon Creek, which runs north around Siffleur Mountain and into the North Saskatchewan River on its south bank, upstream from the mouth of Siffleur River, is named for the same man. In his memo of 15 Jan. 1926, Bridgland stated that "It was due to Prof. Loudon that I came west" but he referred to him erroneously as Professor of Physics at the University of Toronto "since 1900" ("Correspondence related to the 1924 photo-topographical Survey of the Siffleur, Ram & North Saskatdhewan rivers ...," LAC, RG 88, vol. 247, file 18522). James Loudon was professor of physics and chairman of the Department of Physics in 1901, when Bridgland was a full-time student. (There was but one full professor per department at the University of Toronto in those years, and James had been made a professor back in 1875, and is listed as Professor of Physics and Mathematics in the university's calendar for 1883.) By contrast, William James Loudon was made an associate professor only in 1902, after Bridgland's departure, and a full professor in 1908. But James Loudon had been dead for ten years when Bridgland wrote his memo in 1926, and the memo does not refer to the late professor. William James Loudon did not retire until 1930, and had grown well known for his publications, as his uncle never had; moreover, of course William James had been enjoying the rank of professor for twenty-four years by the time that Bridgland wrote his memo. So, even though James Loudon, who likely enjoyed more prestige, had better contacts than his nephew, and was the first Canadian-born head of a department at the University of Toronto, might seem the more likely person on whom the honour of a geographical name might have been bestowed, Bridgland's direct involvement in the nomination suggests that it is named for the historically less memorable Loudon. The mountain and creek went unnamed until 1956, well after the date of Bridgland's own death, when the Geographical Names Board of Canada (est. 1897) made the names official. (I thank Harold Averill, of the University of

Toronto Archives, and Gabrielle Zezulka-Mailloux for their dedicated assistance in trying to sort out this question.) For Mt Loudon, see map on p. 182.

12 C.B. Sissons, "In Memoriam: M.P. Bridgland," *Canadian Alpine Journal* 31 (1948): 219.

13 Robert W. Service, "The Call of the Wild," *The Spell of the Yukon* (New York: Barse and Hopkins, 1907), 30–2.

14 Bridgland's "limited examination" required answers to the following:

1. Write a composition of not less than 200 words on:
    The natural resources of your province.

2. Prove the rule for converting a recurring decimal to a vulgar fraction.

3. If 8 per cent be gained by selling a piece of ground for $4,125.60, what would be gained per cent by selling it for $4,202?

4. Write down the 3rd term of $(a + b)^{15}$.

5. How much ore must one raise, that on losing $17/40$ in roasting and $8/19$ of the residue in smelting, there may result 506 tons of pure metal?

6. Solve $2^{x+1} + 4^x = 80$

7. Construct geometrically the relation $(a + b)\, b = a^2$.

8. Solve $x^2 - 7x + \_(x^2 - 7x + 18) = 24$.

9. Extract the seventh root of !00317 and multiply by $(824)^{2/3}$.

10. Given log. 2 = !301300, log. 3 = !4771213, find log. of !024, 375 and 432.

11. The perpendiculars from the vertices of a triangle to the opposite sides respectively, intersect in a point.

12. Construct a rectangle equal to the differences of two given squares.

13. Show that $\sin A + \sin B = 2 \sin \_ (A + B) \cos \_ (A - B)$.

14. Show that $\tan (A + B + C) + \dfrac{\tan A + \tan B + \tan C - \tan A \tan B \tan C}{1 - \tan A \tan B - \tan A \tan C - \tan B \tan C}$

15. Given A = 30%, B = 45%, c = _ 18; solve the triangle.

16. In a plane triangle the sides are 7, 8, 9; find one of the angles.

17. In a spherical triangle A = 68%, B = 72%, C = 80%; find one of the sides.

18. Give formulæ for volume and surface of a sphere, cylinder, cone, and frustum of a pyramid.

19. How many acres in a field whose sides are 13, 14, and 15 chains?

    (Canada, Department of the Interior, *Report of the Surveyor General*, Sessional Paper No. 25 [Ottawa: Dept. of the Interior, 1903], 99–100.)

15 A.O. Wheeler to É.-G. Deville, 27 Dec. 1901, DLS, Dept. of the Interior, LAC, RG 88, vol. 167, file 1996.

16 A.O. Wheeler, *The Selkirk Range*, 2 vols. (Ottawa: Government Printing Bureau, for the Department of the Interior, 1905), 1:24.

17 It should be noted, in fairness to him, that Wheeler had one permanently injured arm. Having, with fifty others, volunteered to organize the DLS's Intelligence Corps to oppose Louis Riel's Métis forces during the second North West Rebellion in 1885, he suffered a bullet wound in the shoulder.

Wheeler was no more patient as an instructor than as a student. Although Bridgland left no records of his first climb with his chief, his university friend Charles Sissons recalls his first day four years later, in 1906, and shows Wheeler as a challenging chief of party:

> My first day's work on the survey was a time of testing. In fact, I thought then that A.O. Wheeler intended it should be so. The ay was assigned to measuring the movement of ice on the Yoho Glacier. To this end several markers of iron were to be driven into the ice near the snout and the exact location of each determined by instrument and recorded. We began the day by fording the swift-flowing and icy-cold Yoho as it emerged from the cavern at the snout of the glacier. We were supposed to dry and warm ourselves as we worked on the ice-tongue! Then in the evening A.O.W. set off at a terrific pace for home. The pace was continued up a steep slope of a thousand feet or more into camp. By this time my stiffness had disappeared, and I followed him hard on his heels, my spirits rising as I saw the sweat roll down the back of his neck. Nothing was said, but I think he was satisfied with his new hand—rather better than I was with the common sense of the great man. There had been no need whatever to ford that creek. Bridgland used to tell of one new member of an earlier party who failed to stand his test. On his first experience with a glacier he fled.

(C.B. Sissons, *Nil Alienum: the Memoirs of C.B. Sissons* [Toronto: University of Toronto Press, 1964], 79–80.)

18 Esther Fraser, *Wheeler* (Banff: Summerthought, 1978), 115.

19 Janice Sanford Beck, *No Ordinary Woman*, 107.

20 A.O. Wheeler, M.P. Bridgland, and A.J. Campbell, "The Application of Photography." The attribution of co-authorship appears in J. Monroe Thorington, "Members elected after December 1902 and prior to 1 January 1912," *American Alpine Journal* 6 (1947): 348. In fact, it appears that sequential authorship is a more apt description of the collaboration, for the first section sounds as if Wheeler's voice alone is present in the narration, while the second section is sub-titled "Notes by M.P. Bridgland, D.L.S., and A.J. Campbell, D.L.S.," 87–96.

21 With the outbreak of the First World War, reduced budgets and a shortage of staff limited activity in the Ottawa office. Nevertheless, in 1914, with the Grand Trunk Pacific Railway (GTPR) completed through Yellowhead Pass, a survey had to be done before Jasper Park was overtaken by tourists. This was Bridgland's work in 1915. Similarly, despite the war, urgency arose to complete a survey of the Alberta-British Columbia boundary. Regarded as Wheeler's claim to fame, this survey took him and his associates from 1913 to 1924 to complete. (Wheeler was retained by the province of British Columbia, while Alberta, and, in 1915, Canada assigned their interests in the survey to Richard W. Cautley [1873–1953], DLS, another renowned figure in western surveying in the early

twentieth century [John A. Whittaker, compiler, *Early Land Surveyors*, 29–31].)
After the war, the DLS gave the Ottawa office plenty of business by resuming
surveys throughout Canada, and in the 1920s by working with the Canadian
military and the Geological Survey of Canada to develop a uniform system
of maps, which later became known as the National Topographic System.
Mountain and prairie surveys also continued, as did surveys on several provincial
and territorial boundaries.

22 Department of the Interior, *Annual Report* (Ottawa, Dept. of the Interior, 1902),
Sect III, 9.

23 Bliss Carman, "May in the Selkirks," in *Canadian Poetry: From the Beginnings through
the First World War*, selected and with an afterword Carole Gerson and Gwendolyn
Davies (Toronto: McClelland and Stewart, 1994), 244.

24 Lizzie Rummel, interview with Ley Edward Harris, 4 Sept. 1972; Elizabeth
Rummel Fonds, Whyte Museum of the Canadian Rockies, M28/S9/V554.

25 Lizzie Rummel, interview with L.E. Harris, 16 Mar. 1976; Elizabeth Rummel
Fonds, Whyte Museum of the Canadian Rockies, M28/S9/V554.

26 Wheeler, *The Selkirk Range*, 69.

27 Chic Scott, *Pushing the Limits: The Story of Canadian Mountaineering* (Calgary: Rocky
Mountain Books, 2000), 42–3.

28 Wheeler, *The Selkirk Range*, 103.

29 E.O. Wheeler, "In Memoriam," *Canadian Alpine Journal* 31 (1948): 221.

30 See appendices for a chronology of surveys completed by Bridgland or under
his direction.

31 http://www.em.gov.bc.ca/Mining/Geolsurv/Publications/OpenFiles/
OF1992-16-Pioneer/Fossil.html

32 Lillian Stewart, "Burgess Shale Site," *Canadian Encyclopedia*, 2nd ed., 4 vols.
(Edmonton: Hurtig, 1988), 299. Walcott named the shale site Burgess after the
nearby mountain of that name.

33 Blaeberry River and Howse Pass had also been visited by James Hector in early
September 1859 as part of the Palliser survey, in October 1871 during Walter
Moberly's surveys for a railway route through the mountains, and by alpinists
Hugh Stutfield (1858–1929) and Norman Collie (1859–1942) in 1902. (See Spry,
ed., *Papers*, xciii, 443–47; Sleigh, *Walter Moberly*, 178–81; and Hugh E.M. Stutfield
and J. Norman Collie, *Climbs and Exploration in the Canadian Rockies* [New York:
Longmans, Green, 1903], 59–64.)

34 At a dance and reception at a Masonic Lodge, Ina met Frederick Hess and the
young couple later married. Hess likely met his wife at the suggestion of M.P.,
his good friend.

35 Lantern slides were the glass precursors to the plastic-film slides we know today.

36 Margaret Hess was appointed a member of the Order of Canada in 1981 for her
contributions to scholarly research on the arts and crafts of Canadian First
Nations.

37 Bridgland's knowledge of resources could have also played a part in his own investment decisions, as well: he bought shares in a number of British Columbia mines after retiring from survey work. In 1931, when his days with the DLS were over, his son Edgar remembers, father and son visited a blaster gold mine near Barkerville, British Columbia, and stayed in cabins there for about a month. The next year, Edgar and his father went into the mountains again, to a gold mine in which M.P. invested with their next-door neighbour in Calgary, a Mr. Fitzsimmons. This mine was located on Gold Creek, on the Columbia River's Big Bend upriver from Revelstoke. The area was especially familiar to Bridgland, and it was the location of some of Hess's timber limits. Bridgland took on much of the physical work around the mine while Edgar and Hugh, Fitzsimmons' son, would spend time together. For the most part, Bridgland worked the monitor and superintended the sluicing equipment. There was a big fire hose with an eight-foot nozzle, from which water shot down into a screen-bottomed sluice box twenty to thirty ft long into which the heavy gold was dropped in order for the dirt and other rocks to be separated out through a series of different-sized screens.

38 Elizabeth Parker, "The Alpine Club of Canada," *Canadian Alpine Journal* 1 (1907): 3. I thank PearlAnn Reichwein for supplying Parker's dates. Parker served as the club's secretary through 1910. Readers interested in Parker and in both the ACC and women's participation in it should consult Reichwein, "Parker, Elizabeth," *Biographical Dictionary of American and Canadian Naturalists and Environmentalists*, ed. by Keir B. Sterling, *et al.* (Westport, Conn.: Greenwood Press, 1997), 609–10; Reichwein, "At the Foot of the Mountain: Preliminary Thoughts on the Alpine Club of Canada, 1906–1950," in *Changing Parks: The History, Future and Cultural Context of Parks and Heritage Landscapes*, ed. by John S. Marsh and Bruce W. Hodgins (Toronto: Natural Heritage, 1998), 160–76; and Reichwein and Karen Fox, "Margaret Fleming and the Alpine Club of Canada: A Woman's Place in Mountain Leisure and Literature, 1932–1952," *Journal of Canadian Studies* 36.3 (Fall 2001): 35–60; and Reichwein and Fox, eds., *Mountain Diaries: The Alpine Adventures of Margaret Fleming 1929–1980* (Edmonton: Historical Society of Alberta, 2004).

39 *Canadian Alpine Journal* 1 (1907): 3.

40 E.O. Wheeler, "In Memoriam," *Canadian Alpine Journal* 31 (1948): 221.

41 These two Swiss guides, whose services were donated to the camp by the CPR, had originally moved to Canada when the railway hired them to attract tourists to the mountains with the promise of guided climbs with authentic Swiss alpinists.

42 The ACC was the first of the alpine clubs to include women in its membership. From its inception, both sexes comprised the membership and filled offices. On the first climb, the two women, K. McLennan (Toronto) and E.B. Hobbs (Revelstoke), successfully graduated to active membership.

43 C.B. Sissons, "In Memoriam": 219.

44 C.B. Sissons, *Nil Alienum*, 168.

45 The building no longer stands; the ACC club house is now located down the Bow River valley, in Canmore, Alberta.

46 Conrad Kain, *Where the Clouds Can Go*, ed. and supplemented by J. Monroe Thorington (New York: American Alpine Club, 1935).

47 Sissons, *Nil Alienum*, 5.

48 E.O. Wheeler, "In Memoriam": 221.

49 Sissons, "In Memoriam": 219.

50 The Ladies' [*sic*] of the Alpine Club, "An Appreciation," *The Alpine Herald* 1.1 (1907): 1.

51 From his work carried out between 1910 and 1912, Bridgland produced the extensive *Report of the Triangulation of the Railway Belt of British Columbia Between Kootenay and Salmon Arm Bases*, which not only provided technical information about the surveys but also described the land and made suggestions for its best uses. In addition to this publication, he wrote several short articles for the Department of the Interior on the methods of topographical surveying, which it distributed internationally in reply to requests for information on this mapping technique.

52 C.B. Sissons, "In Memoriam": 219–20.

53 M.P. Bridgland, Diary 1911. Provincial Archives of Alberta, 79.27, no. 12305.

54 M.P. Bridgland to É.-G. Deville, Surveyor General, 4 Jan. 1914, LAC RG 88, vol. 345, file 13446. Bridgland herein refers to the upcoming 1914 season's needs as being "of the same strength as last year." Identification of Hyatt as the assistant occurs in M.P. Bridgland to É.-G. Deville, Surveyor General, 14 May 1914, LAC, RG 88, vol. 345, file 13446.

55 Department of the Interior, Topographical Surveys Branch, *General Instructions to Surveyors in charge of Parties for the Survey of Dominion Lands in Manitoba, Saskatchewan, Alberta, the Northwest Territories and the Block of Three and One-Half Million Acres in British Columbia* (Ottawa: 15 Apr. 1913). It appears from a letter written by Bridgland in September 1915 and quoted below that the chief of party's pay increased to $9 per day in 1914.

56 M.P. Bridgland to É.-G. Deville, Surveyor General, 23 Feb. 1914, LAC, RG 88, vol. 345, file 13446. The total of glass plate negatives appears in M.P. Bridgland to É.-G. Deville, Surveyor General, 26 Sept. 1913, LAC, RG 88, vol. 140, file 13446.

57 M.P. Bridgland to É.-G. Deville, Surveyor General, 4 July 1913, LAC, RG 88, vol. 345, file 13446.

58 M.P. Bridgland to É.-G. Deville, Surveyor General, 23 Feb. 1914, LAC, RG 88, vol. 345, file 13446.

59 M.P. Bridgland, "Appendix No. 9, Abstract of the Report of M.P. Bridgland, DLS," n.d. LAC, RG 88, vol. 142, file 13913.

60 M.P. Bridgland, "Abstract of the report of M.P. Bridgland, D.L.S.," no date, LAC, RG 88, vol. 142. file 13913; and Bridgland, Field Book for 1914, LAC, RG 88, DLS, Field Book no. 14071.

61 Bridgland, "Abstract."

{ BUT TOURISM REALLY WAS THE PARAMOUNT
CONCERN NOW THAT TWO RAILWAYS WERE
OPENING THE ATHABASCA VALLEY.

SO IT WAS IMPORTANT
TO COMPLETE THE
SORT OF INVENTORIES
THAT COULD INTEREST
AND RELIABLY INFORM
AN ANTICIPATED TIDE
OF TOURISTS.

THREE

Bridgland's Survey of
Jasper Park 1915

✳

One Last Photograph

On 13 October 1915, M.P. Bridgland stood on a small cliff on the
north side of the Miette River. Facing east towards the town of
Jasper and his base camp, he exposed the final photographic
plate for the survey of Jasper Park that he had been working on
for a full summer. He conducted the survey as part of his normal
summer work for the DLS. This particular summer's work had
required 735 photographic plates.[1] Although an end had finally
come to more than one hundred days of climbing mountains to
establish ninety-two photograph stations, the survey had come
only to the end of the work needed to be done in the mountains
and valleys. Now it was time for Bridgland to trade in his boots,
ice axe, transit, and twenty-pound camera for drafting pencils and
tracing linen. As it had been every summer since 1902, making
a map from all the photographs was the ultimate
goal.

THROUGH THE COURSE

OF HIS CAREER,

THE TEACHER IN

HIM...SURFACED IN

PAMPHLETS AND

ARTICLES OF VALUE

TO BOTH THE

SCIENTIFIC COMMUNITY

AND THE GROWING

NUMBERS OF TOURISTS

WHO VISITED THE

ROCKIES AND

SELKIRKS.

*S*pending almost two full years working in the office of the DLS in the Thomas Block in Calgary, Morris devoted himself to creating from those 735 photographs the first comprehensive topographical maps of Jasper Park, an area in which only a handful of the mountains had been named or accurately located. In conjunction with his accompanying survey notes of them, he would use each of the hundreds of images he had exposed onto glass plate negatives to calculate the height, breadth, and pitch of all the mountains adjacent to the railways in the park and up specific valleys such as Fiddle and Tonquin. After he had made thousands of calculations, he was ready for the painstaking process of drawing contour lines by hand on his map. These would follow the convention in cartography that gives a sense of three dimensions on a flat sheet of paper.

In our review of the careers of McArthur and Wheeler, we have seen that the Mapper of Mountains was neither the first nor the most prominent surveyor to use photogrammetry to create maps, but Bridgland did more of it and at a higher quality than any of his predecessors. And once aerial photography came into being on a regular basis following the Second World War, photogrammetry would become a lost art or science, or whatever combination of the two better describes this demanding work. So Bridgland, already a veteran mapper by the time he came to Jasper, would also have no successors. He was not only the first phototopographical surveyor of Jasper Park but also the only one to survey a large portion of it. By 1915, he was also one of Canada's most accomplished mountaineers, with over sixty first ascents to his credit. Through the course of his career, the teacher in him, which his university yearbook had first identified, surfaced in pamphlets and articles of value to both the scientific community and the growing numbers of tourists who visited the Rockies and Selkirks. As well, his exceptional maps were often the only ones available for much of the mountains' rugged terrain. Although he was first and foremost a surveyor for the DLS, he also made it his personal mission to furnish information that would open the mountains to all who cared to visit them. His maps, the trails he either followed or blazed, the techniques he mastered, the climbing routes he put up, and the land-scapes he surveyed form a legacy that continues to shape the mountains as we know them and enjoy them today. This is nowhere more the case than with Jasper Park.

✳ Creating a Park ✳

MANY FACTORS CONTRIBUTED to the development of what since 1930 we have known as Jasper National Park. While one was the protec-tion of the headwaters of the rivers that could be used to irrigate farms out on the prairies, another was the documentation of mountain passes that served as trade and transit corridors. As many as eight thousand years ago, the presence of bighorn sheep and caribou, and the possi-bility of trade with people from the Pacific Slope, would have attracted hunter-gatherer peoples from the East and the North to the region. In the seventeenth century, the main valleys had been home to Sarcee (Tsuu T'ina) people, most of whom had vacated the area by the time David Thompson passed through in 1811. Throughout the nineteenth century, the area saw much activity from European exploration and trade in furs, leather, and grease. A mixture of European, Shuswap, Iroquois, and Cree populated the area that would become Jasper National Park, travelling through and working as hunters, gatherers, anglers, traders, guides, and, eventually, surveyors.

As we have seen, the history of surveying in Jasper is closely related after the end of the fur trade period (1870s) to the history of the area's railways. With the choice of a route for the CPR going to Bow River valley, Kicking Horse River, and Rogers Pass farther south, the Athabasca valley would remain inaccessible by train until September 1911 when the Grand Trunk Pacific Railway, which had arrived in Edmonton in 1909 after it had become the capital of the new province of Alberta in 1905, reached Fitzhugh, now the town of Jasper, a nine-hour trip from Edmonton.[2] The GTPR offered stiff competition to a third firm in the West, the Canadian Northern Railway (CNoR), and stories circulated of surveyors' attempts to sabotage the competing company's efforts by pulling up their survey stakes.

Jasper Park was established about fifteen years after several individ-uals and families had settled in the mountains along the flats bordering

MANY FACTORS CONTRIBUTED TO THE DEVELOPMENT OF WHAT SINCE 1930 WE HAVE KNOWN AS JASPER NATIONAL PARK.

the upper Athabasca River. It was brought into being first as Jasper Forest Park when *Parliament of Canada Order in Council 1907–1323* was passed on 14 September 1907. The Department of the Interior's Forestry Branch managed the park but the chief superintendent of Parks directed the management. In December 1909, as a result of a joint tour made of the forest park in the fall of 1909 by Robert H. Campbell, superintendent of Forestry at Ottawa, and Howard Douglas (1852–1929), commissioner of Parks at Edmonton, John W. McLaggan was appointed acting superintendent. He was given a staff of two game wardens. In 1911, the Department of the Interior established the Dominion Parks Branch, through the passage on 19 May of the *Dominion Forest Reserves and Parks Act*.[3] James Bernard Harkin (1875–1955) was appointed the branch's first commissioner. In that year, responsibility for the newly named Jasper Park was reassigned, and its first resident park superintendent, Col. Rogers, was appointed in March 1913.[4] While these bureaucratic developments were occurring, a decision was reached that the federal government would own and manage the land in its entirety. Thereby, officials deemed the pioneer homesteaders of the Athabasca valley squatters on Crown land. Six Métis (mixed-blood) families dating from fur trade times named Finley, Moberly, and Joachim, some of Iroquois heritage, their parents or grandparents having come west from the Montreal area to service the fur trade, were told by McLaggan early in 1910 to take the financial compensation offered them for their homes and farms, and leave the valley for good.[5] Not surprisingly, most were reluctant to leave. One of the most reluctant, a seventh homesteader, was also resistant: the lone white man, Lewis Swift (1854–1940), had tried unsuccessfully to file a claim to the land he had developed. When a survey party staked the GTPR line through his cabin, Swift brandished one of his hunting rifles in the name of property rights and held railway workers at gunpoint until surveyors agreed to re-stake the line. He insisted that he was not a squatter; after all, he had applied for patent before the park had been created. A subsequent court ruling in his favour granted him title in 1911. His became the only freehold land in the park. In part because they were Métis, in part because a firearms ban precluded hunting, the other pioneer families were not given the option of legal recourse and were successfully evicted.[6]

Ironically, other and much more intrusive development was permitted. For instance, the park was home to hundreds if not thousands of railway workers in 1911, when lines were building.[7] Coal mining began in the Athabasca valley at Pocahontas in 1908, a year after forest reserve designation had occurred.[8] After 1910, when it became known that the Fiddle River valley had hot springs (now known as Miette Hot Springs) like those at Banff, tourism became the primary economic focus in the Jasper area. The mines closed in 1921 but tourism was on the verge of opening new avenues of possibility, ones that accorded with contemporary ideas of wilderness, as homesteading obviously did not. By 1914, the CNoR, finished laying its track through the park. With the expectation of increased passenger train traffic, the hamlet of Jasper began to grow. Trains did not make it boom yet, however, and automobile tourism had yet to dawn. Six vehicles reached the east gate of the Park in 1926 but it was 1931 before anything regarded widely as a passable road from Edmonton to Jasper was opened, and not until 1959 could cars motor along a highway fully paved from Edmonton to Jasper.[9] Similarly, only on 1 July 1940 did the nearly decade-long construction of the Banff-Jasper Highway come to completion and the highway open to traffic between the two national parks.[10]

The effort at attracting tourism to the new park and what was the very remote settlement of Fitzhugh/Jasper involved placing a splash in the newspapers of eastern Canada and in arranging for a celebrity to visit Jasper. As early as the fall of 1909, it was referred to in *The Globe* (Toronto) as a "new wonderland," and the title of D.J. Benham's splash in the same newspaper's weekend magazine section in January 1910, referred to the park as "Canada's new national playground."[11] Benham, who had travelled with Howard Douglas, the parks commissioner, and Robert Campbell, forestry superintendent, on their tour of inspection in the fall of 1908, noticed that in Jasper Park "there is every diversity of nature to gratify the mountaineer, the explorer, the scientist, the naturalist, or to interest and revivify the sojourner from the plains." In 1914, fifty-five-year-old star novelist Sir Arthur Conan Doyle (1859–1930) and his wife Lady Jean were lured to Jasper by an invitation from the Government of Canada. From 11 to 19 June, they resided in a GTPR railway car at Jasper, there being no hotel available. For part of their stay, the novelist rode a pack train to Tête Jaune Cache and return. Not until the publication in 1924 of *Memories and Adventures* did "To the Rocky Mountains," his account of this trip and one in 1923 appear in book form, but *Cornhill Magazine* published his article-length account

AFTER 1910, WHEN IT BECAME KNOWN THAT THE FIDDLE RIVER VALLEY HAD HOT SPRINGS (NOW KNOWN AS MIETTE HOT SPRINGS) LIKE THOSE AT BANFF, TOURISM BECAME THE PRIMARY ECONOMIC FOCUS IN THE JASPER AREA.

"Western Wanderings" serially between January and April 1915.[12] Meanwhile, of course, the newspapers covered his every move. He was an "outdoor man," he confided to a correspondent for *The Globe* when setting sail from Southampton for New York: "I am looking forward to the camping trip," the paper's front page told its readers on 21 May.[13] Ever the publicist, Harkin, who had indeed seen action in the newspaper world prior to his appointment as parks commissioner, released a special edition booklet in the midst of the Doyles' visit.[14] *The Globe* was impressed:

THE VISIT OF SIR
ARTHUR CONAN DOYLE
TO JASPER PARK HAS
SERVED TO DRAW THE
ATTENTION OF THE
TRAVELLING PUBLIC
OF GREAT BRITAIN TO
THE MAGNIFICENT
ASSET CANADA
POSSESSES IN HER
GREAT NATIONAL
RESERVES.

> The visit of Sir Arthur Conan Doyle to Jasper Park has served to draw the attention of the travelling public of Great Britain to the magnificent asset Canada possesses in her great national reserves. Mr. J.B. Harkin, Commissioner of Dominion Parks, is also carrying on a campaign to make these immense playgrounds better known, and has just issued from the Department of the Interior a most attractive souvenir of the mountain parks. This is a booklet entitled "Just a Sprig of Mountain Heather." Attached to its cover is a piece of real purple heather taken from the alpine meadows of the Canadian Rockies.
>
> It is a novel and very effective way to capture the tourists' interest in Canada's national parks. Mr. Harkin treats in the booklet of the significance of these reserves and of the Dominion's natural heritage of beauty in the form of majestic mountain, peaceful valley and emerald lake. Jasper is the latest of the Dominion parks to be established. It comprises an area of a thousand square miles in the wonderful district recently opened by the construction of the Grand Trunk Pacific through the Yellowhead Pass.[15]

The Doyles were accorded their own Pullman car, "Canada," for their return journey to Jasper, and Doyle did his part, subsequently dubbing Jasper Park

> one of the great national playgrounds and health resorts which the Canadian Government with great wisdom has laid out for the benefit of the citizens. ...[E]verywhere there are the most wonderful excursions, where you sleep at night under the stars

Front Cover of *A Sprig of Mountain Heather* (1914). A real sprig of heather, plucked
from Simpson Pass in what was then Rocky Mountains (Banff) Park,
was inserted in the cover of each copy.

upon the balsamic fir branches which the packer gathers for your
couch. I could not imagine an experience which would be more
likely to give a freshet of vitality when the stream runs thin. For
a week we lived the life of simplicity and nature.[16]

In 1910, Benham had assured his readers that the amenities and aids planned for tourists would include "the completion of a proper topographical survey providing reliable maps and information."[17] A five-year-old publicity campaign was in full swing, then, by the time that Bridgland was called to Jasper. In mid-career, and effectively the star surveyor of alpine terrain, the Mapper of Mountains undertook the assignment that Deville at the DLS and Harkin at the Dominion Parks Branch regarded as their most urgent project.

✳ *More Than a Normal Summer's Work* ✳

IN 1915, from his office in the Thomas Block at 1950–13th Street West in Calgary, Bridgland looked west to the peaks on the horizon. Having spent the previous thirteen summers in them, he had developed a profound passion for the rugged outdoor life of a topographic surveyor. The late arrival of spring made him ever more anxious to be climbing and photographing after a winter of office work. By mid-June he had finished painstakingly plotting control points and delineating contour lines until his maps of the previous summer's work in Waterton Lakes Park, the Crowsnest Forest Reserve, and the Canadian Pacific Railway Belt were complete.

Tourism was not alone in exerting the need for a map. With a recent influx of settlers who planned to farm their land, Alberta, just ten years old in 1915, had grown concerned to ensure its capability to distribute water to arid areas in the southern half of the province. Bridgland's survey of Jasper might prove especially valuable to irrigation projects for it would reveal more of the topography adjacent to the headwaters of the Saskatchewan, Athabasca, Columbia, and Fraser rivers. But tourism really was the paramount concern now that two railways were opening the Athabasca valley. So it was important to complete the sort of inventories that could interest and reliably inform an anticipated tide of tourists. In the course of his photopographical work, he was expected also to survey the quality of timber, the flora and fauna, and the most prominent mountains and areas of particular beauty or interest to visitors, all of which information he was to compile in a guide book for the park once he had completed his map. More than a normal summer's work lay ahead of him.[18]

For the time being, though, Bridgland had to assemble his crew for the summer-long survey. Ideally he would have preferred a party of seven, but with the war demanding more and more men, the chief of party settled for a crew of five: one assistant, two labourers who worked as packers and assistants, and two cooks who would also assist with the pack horses. As he had for the work in the previous two years, he chose the reliable twenty-three-year-old Edward Hyatt, fifteen years his junior. By this point in his career, Hyatt had passed his first set of DLS examinations and been articling with Bridgland for two years after gaining his preliminary certificate.[19] With two sets of instruments, Bridgland and Hyatt could work in different localities with their packers as their climbing assistants, and thus cover twice as much ground. The cooks had many responsibilities apart from meal preparation. These included the care of the horses, chopping wood for the stove and camp fire, gathering brush for bedding, and making repairs when needed to the tents, packs, boxes, and ropes. Bridgland was often demanding of his crew, but never more than he was of himself. As Wheeler had noted, he did not regard camp craft as beneath him and his men respected him all the more for participating in chores. Still, it is no surprise that M.P.'s son remembers his father stating more than once that the most difficult appointment to make each field season was that of cook.

Monday, 21 June 1915 was a fine day when Bridgland mustered his crew in Calgary for the long trip to Jasper. For a fare of $14.10, each man boarded the train for the twenty-hour trip by way of Edmonton. From their point of departure, the Canadian Pacific Railway transported the men north to Strathcona, the town on the south bank of the North Saskatchewan River that had amalgamated with the city of Edmonton just three years earlier. The impressive edifices of the Alberta Provincial Legislature and the GTPR's nearly completed Hotel Macdonald dominated the otherwise level skyline on the north side of the river. The men crossed over and connected to the GTPR train to Jasper.

Bridgland did not set out at the same time as his crew but stayed in Calgary for a couple of days preparing his photographic equipment and testing the speed of the glass plates. As he noted in his compendious booklet about photographic surveying, "before commencing work it is customary to expose a number of plates giving different exposures under conditions as nearly as possible to field conditions. These plates receive normal development and are carefully examined to ascertain which has given the best result.... If in doubt, it is always better to over-

Photograph by T. Smith showing a few of Bridgland's glass plate negatives and their sleeves resting on an underlit light table, Gatineau Preservation Centre, Library and Archives Canada, 2004. Bridgland's glass plate negatives measured 11.4 x 16.5 cm (4.5" x 6.5"). Courtesy Rocky Mountain Repeat Photography Project, University of Alberta, University of Victoria.

expose slightly than to underexpose."[20] Although the method of making glass plate negatives had been standardized, emulsion sensitivity was subject to variations because of age and because of the conditions in which the negatives were stored. As a consequence, every batch of plates had to be tested at different exposure times to determine the sensitivity of the emulsion. For the panoramas that Bridgland was to take from mountain tops, he used Wratten and Wainwright Panchromatic Plates, which offered enhanced breadth of exposure times. With these he used panchromatic negatives, which were sensitive to all wavelengths of light, the yellow and green ranges as well as the actinic (blue, violet, and ultra-violet) ones. Because the plates were sensitive to actinic rays and because

Photograph by H.J. Green of the rock cut at the foot of Roche Miette's Syncline Ridge, Grand Trunk Pacific
Railway line, Jasper Park, c.1911. Courtesy Glenbow Archives NA–915–15.

these rays, which literally make the sky blue to the eye, also obscure
details in the landscape and magnify that obscuring effect over distance,
Wratten and Wainwright orange-yellow G-filters had to be used from
mountain peaks. Thereby Bridgland had found through years of trials
that the effects of atmospheric haze could be diminished if not eliminated. The only drawback to these plates was that, since areas in shade
are lighted by refracted light in the atmosphere, the G-filter intensified
the darkness of the shadows, allowing little detail in dark areas. Because
of the vast range of scenery, from dark pines in deep shade to glaringly
bright snow, alpine scenery was especially difficult to capture well at this
stage of photography.

   His tests completed, Bridgland caught the train to Edmonton late
on Wednesday, 23 June. At dawn, the GTPR train was somewhere near
present-day Hinton, chugging slowly along through forests, grasslands,
and wetlands, dotted with small settlements no bigger than a few dozen
people. Then the line fell in with the Athabasca River, and the smoking
black locomotive pulled itself west and south up its valley towards
Jasper. Looking out the window, Bridgland could see through the

THEREBY BRIDGLAND
HAD FOUND THROUGH
YEARS OF TRIALS
THAT THE EFFECTS OF
ATMOSPHERIC HAZE
COULD BE DIMINISHED
IF NOT ELIMINATED.

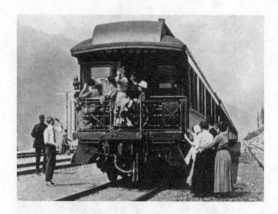

All Aboard! at Jasper Station. Photograph by William James Topley Studio, Ottawa. From M.P. Bridgland, *Description of & Guide to Jasper Park*, 26.

drizzly morning light two bare rocky peaks, Roche à Perdrix on the east side of the tracks, and Bedson Ridge and Black Cat Mountain on the west. Passing between them, the trained entered Jasper Park in the pouring rain on Thursday morning. There stood Roche Miette, by 1915 as well-known a landmark in the Canadian Rockies as any other single peak (now with a passage dynamited for the GTPR through its shoulder by the river). Still, to the experienced mountaineer the first impressions of Jasper and vicinity were a little disappointing. Peaks barely 7,000 ft (2,150 m) high would not have impressed a man who had spent a number of summers in the Railway Belt's Main Range and the Selkirks, but he soon found that climbs even to the lower peaks often involved as much as 4,000 ft to 5,000 ft (1,200 to 1,500 m) of elevation change.

In the rising morning light, the train passed the evidence left behind by the surveyors who made the railway possible: every ten or twelve km along the track there were abandoned log bunkhouses that had been quickly built by and for the railway surveyors, clearing and blasting crews, track layers, and trestle builders. As many as two thousand men had been working in the valley as recently as the previous year. And there was Pocahontas where more than fifty families lived. The town was booming because of the wartime demand for coal. It was the only place in the park serviced with running water and electricity; it also boasted a post office and mounted police barracks.[21]

While some outfitters had also located in Pocahontas, most of them had their headquarters in Jasper, forty km upstream, in the heart of the park at the confluence of the Miette, Athabasca, and Maligne rivers, where most tourists wanted to travel. An hour after leaving Pocahontas, the train pulled into Jasper station. Descending from it, Bridgland was greeted by the site of Jasper's main street. It didn't offer much of a greeting. About two dozen small structures, some log cabins, others simple wood-frame buildings, lined up on a wide dirt road facing the railway lines along the north bank of the Athabasca.

Established twenty years after Rocky Mountains Park (now Banff), the town was beginning to serve as the destination of choice for people seeking "authentic" wilderness experiences rather than comforts such as those offered by Banff and Lake Louise to the south. While Banff had already developed hotels, restaurants, and other infrastructure, the primary businesses in the town of Jasper were geared towards the supply of outfitters. There were blacksmith shops, general stores, a butcher shop, and a veterinary centre. But a restaurant, a couple of cafés, and a large pool hall also operated. Lady Jean Doyle had likely taken her meals in her railway car. The roads were not paved and neither electricity nor running water were available to the townspeople. Although Hugh Matheson had surveyed and sectioned Jasper into lots that could be leased by future investors and settlers, it was a railway hamlet. The permanent population in 1914 was only about 125. It was doubtless slightly greater the next year when Bridgland hit town.[22]

Jasper Park Lodge would not open for another eight years, and no building existed large enough to accommodate all the area's itinerant residents, railway and construction workers, surveyors, mountaineers, scientists, and tourists who arrived on the trains from Edmonton. To solve this problem in the meantime, Robert Kenneth, of Edmonton Tent and Mattress Company, erected a network of ten platform tents between Athabasca River and Lac Beauvert, on the shores of which Jasper Park Lodge reposes.[23] Fabricated from waterproofed canvas, this instant hamlet had been a hallmark of western development for several decades; Dominion Land Surveyors and railway builders alike had thrown up tent cities as their work proceeded west in the 1880s. Aimed at the tourist rather than the labourer, this tent city was the innovation of the GTPR (a dining facility was added in 1919, when the experiment resumed operation after the First World War, and several log cabins were built two years later).[24] In the absence of hotels, still too expensive for the likely return on investment, especially in wartime, the GTPR and Edmonton Tent and Mattress wanted to corner the market for a genuine wilderness experience. (No campgrounds opened in the park before 1927.[25]) One could rent a large tent, complete with brush for a mattress, for a daily fee of $2.50. Staked in the ground in a large clearing was a long wooden sign with "JASPER HOUSE" painted in bold black letters. Tents, each approximately twenty feet square, stood in rows. Beside the main office tent, a sign reading "EAT AND

SLEEP," as well as the day's menu of pig's feet and sauerkraut, identified a large kitchen tent. Having their own tents and cooks, Bridgland's men would not have considered for a moment paying for such a "luxury." Instead, when it opened on 15 June, Tent City's lodgers in 1915 included tourists from as far away as Toronto, Ottawa, New York, Washington, DC, and Seattle.[26]

✳ *First Peak of the Season—Signal Mountain* ✳

M.P.'S PRIORITIES
WERE THE PERFECT
CALIBRATION OF THE
LEVELS OF HIS AND
HYATT'S CAMERAS,
AND VERIFICATION OF
THE FOCAL LENGTH
OF THE LENS.

CONTINUOUS RAIN PREVENTED THE MEN from making any exploratory hikes during their first days in the park. Their routine was familiar on days off. M.P.'s priorities were the perfect calibration of the levels of his and Hyatt's cameras, and verification of the focal length of the lens. He performed this calibration through a series of practical experiments in which he took pictures of targets placed on a level horizon and then applied a set of theoretical equations to the images produced. These tests ensured that he had accurate data at summer's end when he returned to his Calgary office for the cartographic stage of the work. He and Hyatt also overhauled their survey instruments, checking his tripod, transit, and camera. Meanwhile, his packers and cook would have been busy ensuring that the packs carrying clothing, bedding, and personal sundries as well as the wooden boxes expertly crammed full with provisions and cooking equipment were neatly organized and ready for the survey's twelve pack horses when they arrived from Lundbreck, in southern Alberta, where they had wintered after the season in Waterton. The only boxes not entrusted to the beasts of burden were those containing the camera equipment and transit, which Bridgland and Hyatt carried for their respective parties. Because of the instruments' fragility and sensitivity to heat and humidity, they took sole responsibility for them. As Bridgland takes care to point out in his manual,

> the care of plates in the field, both before and after exposure, must be one of the first considerations under all circumstances. Plates are usually shipped in cases containing twenty dozen or more, and, if possible, are carried in these cases throughout the season. They are kept in the tent, though no doubt it would be better, if possible, to keep them where they would be less exposed to variations of temperature. In any event it is

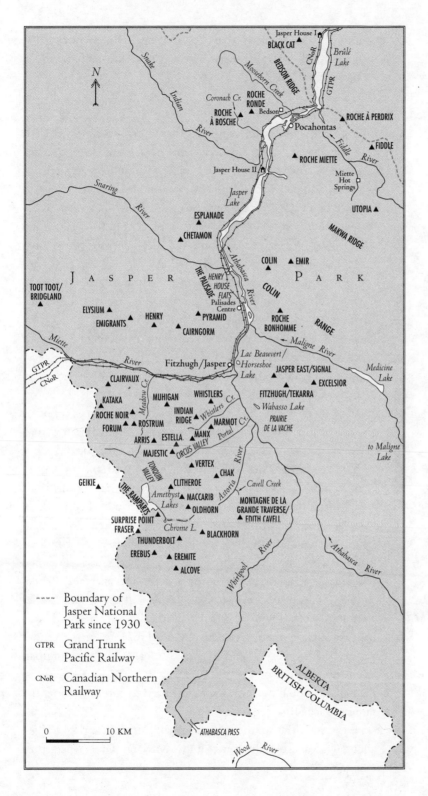

Bridgland's Survey
(with Hyatt) of the
Central Region of
Jasper Park, 1915

absolutely necessary that they be kept dry. When moving, whether by pack train or otherwise, the plates must be handled with the greatest care.[27]

Although he left much of his stock of glass plate negatives and developing equipment at his main camp, for the entire summer M.P. slept beside the camera equipment, including a case of negatives, keeping it dry and out of direct sunlight. Doubtless, Hyatt had adopted the practice.

The horses, shipped by rail, reached Jasper on Friday night. They had been confined to a train car for almost two days. With no corral available

Maligne Gorge. Photograph by William James Topley Studio, Ottawa. From M.P. Bridgland, *Description of & Guide to Jasper Park*, 63.

to pen them in, they managed to make up for their long confinement. While the surveyors slept, they wandered off. A long, tiring search and round-up occupied the men all day Saturday, another soaking wet day. This was just the first of many tricks the horses would play on the surveyors over the course of the summer. They preferred the age-old and infuriating game of filling their lungs with air when the packer was trying to secure a load with ropes and difficult knots. Just as soon as the diamond hitch had been tied, the horse would exhale, leaving the pack loose enough to be shaken off by rearing or brushing up against a tree. If the horses failed to shed their packs by this method, they sometimes connived to have them torn off by running headlong through narrow gaps between trees, wide enough for them but not for the packs.

On Sunday 27 June the rain finally stopped and the entire team climbed its first peak of the season. Perhaps wanting to avoid intimidating his newer recruits with a difficult climb, Bridgland led them from Jasper to the peak of Signal Mountain (7,400 ft [2,255 m]) and back. This relatively easy trip could be made in less than a day, either on foot or on horseback. The men donned the regular surveyor's garb: tweed breeches and flannel shirts, spiral puttees over socks or heavy golf hose, and leather climbing boots. The leather soles of the boots were fixed with nails resembling upholstery tacks intended for extra grip on snow, ice, or other slippery terrain. There was no snow on the mountain but, after a winter out of the field, the men regained a sense of the nails' gripping capacity and of the need to add or subtract the number they affixed to each boot.

East along the southern bank of the Athabasca, they walked about ten km from the town site to Maligne River, where they enjoyed its spectacular canyon in which the river disappears before reappearing and widening out downstream. Jesuit missionary Pierre-Jean de Smet (1801–73) named the river Maligne, meaning "wicked" in French, because he found it difficult to cross in 1846; subsequent travellers concurred and the name stuck. Bridgland's team did not have to attempt a fording. Just as well, for the men's cargo was precious.

The pack with the camera and glass plate negatives was the heaviest and most fragile. On this united outing of the two parties, Hyatt carried the camera and Bridgland carried the transit, which had been shipped from England especially for the DLS by Troughton and Simms of London. The principal parts of it were made of aluminum to keep its weight down to a mere fifteen pounds, but because it was lightweight it could be bent out of shape relatively easily. Bridgland insisted on carrying it because the slightest rough treatment could result in a misalignment in the instrument's mechanisms. Even if the damage was noted before it was used, a great deal of time and energy would have to be spent to re-calibrate it. Failure to notice any misalignment or other damage if it occurred would translate into significant errors in both the readings taken and the final map. Bridgland secured the transit pack to his back with thick leather shoulder straps. The base of the pack was a rectangular canvas sack containing the legs for the tripod; a separate case on top held the transit box, tripod head, and camera base. Hyatt took the thirty-five-pound camera in its mahogany outer box, fitted into a leather case.

THE PACK WITH THE CAMERA AND GLASS PLATE NEGATIVES WAS THE HEAVIEST AND MOST FRAGILE.

A post for the government telephone line that led to the summit of Signal Mountain marked the start of the trail they followed part way up the canyon.[28] Although it was unusual to see caribou in the lower valleys, the party saw a few on the lower benches in the early morning. As the men ascended through brûlé, windfall, and small jack pine, the Athabasca valley hove into view. Facing the surveyors on the opposite side of the valley, Pyramid and Patricia lakes nestled at the foot of Pyramid Mountain; their point of view was higher but similar to the one that Group of Seven painter A.Y. Jackson would adopt about fifteen years later to depict Pyramid Mountain. Directly below, the climbers could see the unmistakable light green to deep blue hues of spring-fed lakes Edith and Annette, as yet undeveloped and unaccompanied by Jasper Park Lodge's golf course, completed only a decade later at a cost of $190,000.[29] Then, gaining the 7,397-foot (2,254-m) summit, they enjoyed the sight of large expanses of the main river valleys where

Photograph by M.P. Bridgland of the Athabasca River valley from Pyramid Mtn, Stn 56, no 456, direction southeast, 1915 (detail). Lac Beauvert or Horseshoe Lake is on the right, Edith and Annette on the left. Courtesy Jasper National Park.

HE LED BY EXAMPLE,
BUT HIS EXAMPLE
WAS DAUNTING.

they would survey that summer. The upper Athabasca would lead them south as far as the mouth of Whirlpool River, and east as far as Pocahontas, while the Miette River valley would direct them gently up to Yellowhead Pass, at the boundary of Jasper (National) and Mount Robson Provincial parks, the latter having been established only two years before (1913). Although they could not see all the tributary rivers they would explore that summer, the sight surely indicated that a formidable task lay ahead of them, one that could not reasonably be accomplished by any other method of survey. But it was already the end of June!

For those who had not climbed before, this day offered the best and perhaps only opportunity to gain mountaineering instruction from Bridgland. He was an accomplished and enthusiastic teacher to many of the novice alpinists he guided during other summers at the Alpine Club of Canada. Some of the patience he needed for that voluntary work was evident in his treatment of novice surveyors. Still, he was a firm believer that experience itself was the best teacher. He might have been deemed by some a demanding person to work for but not because of a flaw in his personality; rather, his stout, robust body was suited to the work. In combination with tremendous stamina, these attributes made him a great hiker who could endure long days. He led by example, but his example was daunting.

However, when it came to surveying, Bridgland knew that every novice needed instruction in basic techniques, and that such instruction early on in the season would save time all summer long and expedite the

Photograph by M.P. Bridgland of the Athabasca River and Miette River valleys from Signal Mtn (with Jasper town site), Stn 27, no 225, direction west, 1915. The crew returned to Signal Mtn in early August to shoot the series of photographs that included this one. Courtesy Jasper National Park.

collection of accurate data. Still, the party would have climbed quickly because Bridgland knew that the best light for photography occurred at about mid-morning. He may well have used this climb as an opportunity to impart to the newcomers strategies for alpine photography, one of which was to "[a]void taking photographs directly under [into] the sun, if possible, particularly if the sun is low. This may be avoided to some extent by taking the views to the right of the sun on first reaching the station and the views on the left just before leaving."[30]

When they reached the easy summit, the men took off their packs and explored the ample mountaintop. Bridgland needed to capture as much of the 360° as he could from each station. Ideally, he could do so from just one, but it was seldom the case that the tops of mountains cooperated. He often was limited by large rocks in the foreground, steep ridges, or rocky promontories that hid parts of the skyline or a

STRONG AND SUDDEN
GUSTS ARE COMMON-
PLACE AT HIGHER
ELEVATIONS, AND
COULD EASILY KNOCK
OVER THE FOURTEEN-
POUND CAMERA OR
MAKE IT SHAKE AND
THUS RUIN THE
EXPOSURE.

major portion of the distant landscape. The key then lay in mini-
mizing the number of stations needed to accomplish a comprehensive
set of plates. In the case of Signal Mountain, two stations were
needed.[31]

First, he set up on the northeast side of the peak. As he had often
in the past and would over and over again all summer, he then took the
tripod legs out of their canvas carrying case and the base out of the
transit box, extending the legs and affixing the tripod to the base. Both
the transit and the camera had three adjustable foot screws, so the same
tripod could be used for both instruments. It had been designed short
by Surveyor General Deville specifically because of the dangers and
elusiveness of stability in mountain work.[32] Even when there was no
wind blowing, Bridgland took great care to stabilize the tripod before
he mounted the camera. He gathered rocks in a small bag, which he
hung from the base, then used rocks and stones to block the bag of
stones and each tripod leg. Strong and sudden gusts are commonplace
at higher elevations, and could easily knock over the fourteen-pound
camera or make it shake and thus ruin the exposure. Hyatt assisted him
as the rest of the crew watched and learned.

✳ *Bridgland's Photographic Methods* ✳

AS USUAL, the camera was a simple metal box, with a Tessar brand
lens and a set of levels, one horizontal, one vertical, welded to it. This
box fit snugly into a mahogany case, with holes drilled to accommodate
the lens and permit reading of the levels. But before levelling the
camera Bridgland had several steps to complete. A folding sun shade,
which was stored in the cover of the leather camera case, had to be
hooked on the front of the camera to shield the lens from any direct
sunlight. The leather case also contained a tin box holding twelve metal
plate holders, each with an unexposed glass plate negative measuring
4.5" x 6.5" (11.4 x 16.5 cm). He extracted a holder and placed it in the
"carrier" at the back of the camera, then turned the camera in towards
the Athabasca River and Jasper town site for his first photograph of the
season. The tripod stood only 3.3' (100 cm) high when fully extended so
he had to kneel to focus the view.

Unlike today's sophisticated 35-mm or digital cameras, Bridgland's
required much practice and even more patience. Perhaps the most deli-
cate task before exposing the plate was levelling the camera by means of

three "foot screws" at the base. These he adjusted until the bubbles centred in the levels that he could check through the mahogany camera box. Although levelling was fairly straightforward, it was complicated by the fact that the horizon line of the landscape being photographed had to pass through an exact point on the negative in order to permit use of the view at the map-making stage to calculate distances and angles of various features in the landscape. While testing his equipment in Calgary, he had taken a small, sharp file and made four precise notches inside the camera box. These were set into a metal rim that fit next to the glass plate negative while it was being exposed, protecting the border from exposure. When he levelled a view, the horizon had to pass through the corresponding horizon line notches that he had marked on the mahogany case of the camera, and the bubble in the level had to be centred.[33] As if this was not already complicated enough, when Bridgland knelt to focus his view he did not see the image of what he was about to photograph but rather this image in reverse and upside down! Obviously, such work was painstaking, not at all suited to someone with an impetuous, careless, or unmethodical disposition, and seldom straightforward in the vicissitudes of mountaintop terrain and weather.

OBVIOUSLY, SUCH WORK WAS PAINSTAKING, NOT AT ALL SUITED TO SOMEONE WITH AN IMPETUOUS, CARELESS, OR UNMETHODICAL DISPOSITION, AND SELDOM STRAIGHT-FORWARD IN THE VICISSITUDES OF MOUNTAINTOP TERRAIN AND WEATHER

   With the view lined up and the camera ready to go, Bridgland removed the slide (a thin piece of black, rigid plastic that fit between the negative and the plate holder), checked the levels one last time, and then carefully removed the lens cap to expose the plate. It was his policy to use a small stop, not larger than $f/32$, throughout the season, and note the time exposure in the transit notes. The camera's stability was vital in part because exposure times for the glass plate negatives varied from a half-second to as much as three seconds in different light conditions. Any movement during these lengthy exposures would have resulted in a blurring on the plate. As Bridgland's images prove, he always sufficiently secured the camera to prevent this result or else, having noticed movement, took another exposure before leaving the mountain to ensure that he had a replacement. Once he had his exposure, he replaced the slide, and then withdrew the plate holder, returning it to its designated slot in the camera pack's tin box. He took several photographs in this manner: "The general practice is to expose the plates from left to right, allowing a generous overlap between each plate for the sake of contiguity. This overlapping is also done to eliminate, in so far as is possible, the distortion in the print, which has been found, in practice, to be greatest at the edges."[34] Then he moved his camera to a second station, on the southwest side of the peak, where the chore of securing the tripod began again. From here he finished photographing a full circuit of the horizon. As he moved between stations he did what he had done every previous year: he measured his distance from the place where the transit readings were made, usually the first station, and noted the information in a department-issued notebook. This 3" x 5" (7.6 x 12.7 cm) hardbound book of graph paper was marked "Transit Notes." In it, as well, he noted not only the location of all camera and transit stations but also any angles of elevation or depression between himself and the transit site. Each angle or other measurement that he noted would help him back at the office when drafting the map. Thereby over the course of the summer he gradually amassed the second most valuable collection of data after the photos themselves.

The essential instruments for phototopographic surveying are the transit—or theodolite—and the camera. The transit is used to measure vertical and horizontal angles by means of a telescope that rotates on two axes along graduated scales. As they rotate the telescope, surveyors can determine the relative position of landmarks according to brass circles three inches (7.6 cm) in diameter, etched with lines marking minutes (units of measure equal to one-sixtieth of one degree) by which angles between prominent landmarks could be gauged. First, Bridgland adjusted the three screws on the base to level the transit. Then he determined the angles for at least one well-defined point in each photograph and preferably two or more. If the points were not either cairns at stations that he and his men had built earlier in the survey or well known peaks, he would make a sketch of the peak's shape in his transit notebook so that he could recognize and re-identify the corresponding data and photographs during the map-making process. He calculated the angles from a central transit site, working around any obstructions. On each station he and "the boys," as he called his assistants and packers, built a cairn that was big enough to be seen easily from other summits or points in the valley. In forested areas where cairns were impractical, a flag of black or white cloth was tied to an exposed branch to serve as a marker, but on this first peak the men gathered enough rocks and scree to build a cairn about six ft (two m) high.

ON EACH STATION HE AND "THE BOYS," AS HE CALLED HIS ASSISTANTS AND PACKERS, BUILT A CAIRN THAT WAS BIG ENOUGH TO BE SEEN EASILY FROM OTHER SUMMITS OR POINTS IN THE VALLEY.

Cairn built by M.P. Bridlgand's survey crew on Mt Emir, Stn 42, in 1915. Location Photograph by E.S. Higgs, 1999.
Courtesy Rocky Mountain Repeat Photography Project, University of Alberta, University of Victoria.

✳ *More Peaks and Valleys* ✳

SIGNAL MOUNTAIN had not officially been named in 1915, so Bridgland simply called it "Jasper East" in the survey diary in which he recorded notes about the weather and the stations his and Hyatt's teams occupied. He later submitted to the surveyor general the suggestion that the peak be called by the name it bears today. During that first climb, he had scouted a wonderfully flat area near good water and plenty of wood, just above the tree line. The following day he and Hyatt split up, Bridgland moving his men, horses, and supplies back up Signal to set up one camp while Hyatt took two men with him to the north side of the Athabasca River to set up a second.

On the morning of Wednesday, 30 June, Bridgland summitted Mt Tekarra (8,839 ft [2,694 m]), which he called Mt Fitzhugh, and which lay three hours away by horse south of camp. As Hyatt was now off

Photograph by M.P. Bridgland of the Jasper town site and Athabasca River from Old Fort Point, Stn 26, no 215, direction west, 1915. Courtesy Library and Archives Canada R214.

with his own crew, M.P. worked with J.W. Martin, who carried the camera and assisted with the photography, the cairn building, and the transit readings.[35] The next day turned too cloudy for any photography in the morning, but when the skies cleared in the afternoon Bridgland completed a survey from Mt Excelsior (9,003 ft [2,744 m]), a five-hour climb from camp, then moved down to Maligne Canyon, which he later deemed "one of the finest gorges to be found in the Rocky mountains" and "one of the most important scenic attractions in the vicinity of Jasper."[36] On his fourth day of climbing, he ascended Roche Bonhomme (8,186 ft [2,495 m]) in the Colin Range, across the Maligne from Signal and downstream on the Athabasca River. Then, in the late evening's pouring rain, he returned to the main camp near the town site.

With the showers persisting the following morning, he attempted no heavy climbing, but made the short hike to Old Fort Point across the river from the town site. Although this point is but a bare knoll about 400 ft (120 m) high, it offers a splendid view of one stretch of the Athabasca valley and remains today a popular viewpoint. The principal difference between Bridgland's view and a modern one is that where Jasper Park Lodge stands today the tent city stood on the shore of Lac Beauvert. The lake would soon be the scene of great tragedy. For his part during this first short week of work, Hyatt and E. Norfolk, his assistant, made trips to the summits of Pyramid Mtn (9,075 ft [2,766 m]) and Mt Cairngorm (8,563 ft [2,610 m]), respectively, then returned to the main camp on Saturday, 3 July. Sunday was deemed a day of rest for all but the chief who decided to spend another wet day developing some plates to check for problems with the first week's work.

✳ *Inside Bridgland's Dark-tent* ✳

FREQUENTLY IF NOT DAILY, M.P. would set up the small developing outfit he carried with him. The dark-tent was used daily to change plates and occasionally to develop some of them. (The development of photos in the field was necessary for the detection of problems, if any, and the correction of them before an entire season's work was wasted.) Made of two thicknesses of black sateen and one of red, the dark-tent had no distinct shape and was either suspended from the top of a camp tent or placed over three or four short stakes driven into the ground from which the material hung low enough that the bottom could be tucked in to exclude all light. M.P. advocated carrying two dozen plate holders per camera so that negatives could be changed by the dozen. He did all he could to avoid the confusion of half boxes of exposed plates and the possibility of double exposures. As the negatives were carried in light-proof plate holders, he could take the full dozen out of the tin box inside the camera case and replace them immediately with a dozen pre-prepared unexposed plates. Into the dark-tent went the exposed plates, still in their holders, as well as a wooden case containing boxes of negatives. Thereafter, he mixed and set up all the chemicals he needed for their development.[37]

INTO THE DARK-TENT WENT THE EXPOSED PLATES, STILL IN THEIR HOLDERS, AS WELL AS A WOODEN CASE CONTAINING BOXES OF NEGATIVES. THEREAFTER, HE MIXED AND SET UP ALL THE CHEMICALS HE NEEDED FOR THEIR DEVELOPMENT.

The air was stuffy in the heavy silk tent, sealed off as it was. With deliberate, patient movements in the complete darkness, M.P. felt for and unbuckled the leather straps that secured the top to the case of negatives. Reaching for the plate holders he had positioned on the ground beside him, he removed the exposed negatives and placed them in the top of the box in pairs, face to face. When all twelve plate holders were empty, he felt for a box of a dozen fresh plates from the large case, opened it, and began refilling the holders. Blindly, he patiently pencilled the number of the holder and the number of the dozen on the corner of each unexposed plate so that he could keep track of the correspondence between each exposure and his transit notes. Except for the few he set aside for developing, the exposed plates then went into the box just emptied. He was a stickler for repacking the negatives exactly as the maker had shipped them because most of them spent as many as three months in the field before being transported east all the way to Ottawa to be developed. He could not afford even inadvertently to squander the investment of a season's toil and, indeed, the status of his employment.

After placing the negatives in developing trays, covering them with a series of chemicals he had pre-mixed into glass bottles with glass stoppers, and then dusting the plates with powdered chrome alum, he could emerge from the darkness to inspect some of the work he and Hyatt had accomplished. Inspecting Hyatt's plates also enabled him to determine which ranges or mountains were covered between their two sets of photos and which needed to be climbed next. Many considerations bore on the decision:

> When in the field, it is important to select those points which give the best views of the surrounding country. This does not mean that the highest peaks are always the best. In photographs taken from a very high peak, the surrounding country often appears dwarfed, and the details [such as smaller crags and valleys] do not show up as well as in ones taken from a more moderate elevation. It must also be remembered that the higher the peak, the longer the ascent is likely to take, and the greater the likelihood of encountering sudden storms. Frequently very useful views may be obtained with little trouble from points of comparatively low altitude.[38]

Furthermore, every part of the country to be mapped had to be photographed from at least two angles. These had to differ sufficiently from one another to accentuate different shapes in the land, but not so much as to preclude easy recognition of the same landform in two photos. In country presenting broken ridges, distinctly-shaped peaks,

BLINDLY, HE PATIENTLY PENCILLED THE NUMBER OF THE HOLDER AND THE NUMBER OF THE DOZEN ON THE CORNER OF EACH UNEXPOSED PLATE SO THAT HE COULD KEEP TRACK OF THE CORRESPONDENCE BETWEEN EACH EXPOSURE AND HIS TRANSIT NOTES.

glaciers, or large snow patches, M.P. and Hyatt both made pairs of photos at large angles of intersection, varying from ninety degrees to thirty degrees, but in the foothills or areas without boldly distinctive features they used intersections of between thirty and fifteen degrees. M.P. would later be able to calculate the angles and differences of height between most of the prominent features in the pairs of views, and then translate this information into contour lines on a sketch map. Doubtless, the phototopographer-cartographer had honed his methods over many years' experience.

✳ *Surveying to the South and North* ✳

AFTER DISCUSSING HIS PLAN OF ATTACK WITH HYATT, Bridgland decided that they should extend the survey farther south than the range assigned them in their official mandate. The Department of the Interior was primarily concerned with having a map of the area adjacent to the railway but, because of the growing interest in the potential for irrigation projects that could sustain agricultural development east of the Rockies, it was also keen to know more about watersheds. No doubt, Morris was also prompted by tourism's appetite for access to La Montagne de la Grande Traverse (which would be renamed Mt Edith Cavell in 1916), the Tonquin Valley, and perhaps even Athabasca Falls, access to all of which lay up the Athabasca valley past where the railways left it and headed west to Yellowhead Pass. From climbs that he had already made he had seen that the mighty Athabasca flowed from the south, as did the Whirlpool River, its main mountain tributary. In his opinion, "the omission of the heads of the streams in this vicinity would have left the map incomplete and comparatively little extra work was required to include them."[39] So he determined to proceed up the Athabasca to the mouth of the Whirlpool, as well as up the Miette to the vicinity of the height of land at Yellowhead Pass.[40] Having two teams permitted him this option.

Leaving one of his men at the main camp to watch over their huge stock of extra photographic plates and miscellaneous supplies, the other six men and all the horses began together, heading up the Athabasca River valley. Starting on the old wagon road that took them across Miette River, the men turned south on a trail running along the west bank of the Athabasca for about six km, and then onto "an old

Indian hunting trail [that] has not been spoiled by too much work";[41] that is, it was a good trail that, despite the constant rain, had not been reduced to mud through overuse. It took them west and cut across a grassy flat at the base of the Whistlers, two peaks named for the shrill call made by the hoary marmots residing in abundance on their slopes. Following Whistlers Creek to its head, Bridgland established a camp there, while Hyatt continued with Norfolk as well as his cook and horses farther south up the Athabasca before camping along another tributary, Astoria River. After a good night's sleep at a temporary campsite, Bridgland made easy work of climbing Marmot Mtn (8,557 ft [2,608 m]) on Tuesday, 6 July. This peak is much like the Whistlers in both its most visible inhabitants and its rocky upper valleys, in which they make their dens.

After his early morning climb, Morris returned to the campsite where he found everything packed up and his men and horses ready to move on. The pack train journeyed into the uncharted territory towards the Tonquin Valley by way of a branch of the hunting trail they had followed the day before. Winding up through grassy slopes to the 7,500 ft- (2,280 m-) high Marmot Pass, he and his crew then descended 1,500 feet (450 m) of slopes covered with fallen timber down to the banks of Portal Creek. This required much stopping and starting as the men tried to keep ahead of the pack train, cutting through the fallen trees so the horses could pass on the steep slope without losing their footing. Even in their state of concentration and physical exertion, they could not help but notice the beauty of their surroundings, especially such picturesque sights as a point where a branch of Portal Creek emerges from an enormous rock slide originating in Circus valley. Bridgland was tempted to venture into this valley, for, although forbiddingly rocky and barren, it led to Manx Peak (9,987 ft [1,600 m]). The shortest of the three mountains that stand at the north head looked as though it would offer an interesting climb; however, since the only access lies over a long shale slope the traverse of which would doubtless prove very time consuming, he buried his mountaineer's curiosity and diligently carried on towards the head of the creek. The crew established a camp about five km above its forks, just at the timber line. As soon as the horses were freed from their heavy packs, they were brought back down to the luxurious grass of the meadows. As they feasted, three caribou passed close by, obviously in search of similar grazing, but the horses' attention was on the buffet before them.

Portal Creek valley was like a dream for Bridgland in that it offered him the challenge of five as-yet-unclimbed peaks. Challenge enough, it increased dramatically when rain and snow began falling. Undaunted by the weather, he and Martin packed up the transit and camera and headed out to attempt their first as-yet-unascended peak of the season, Mt Maccarib (9,019 ft [2,749 m]). It is somewhat misleading to say that a peak was any less difficult to ascend if it had been climbed before— occasionally, even with his connections to the ACC, Bridgland was not privy to the climbing notes of the mountaineers who had preceded him. Consequently, he had to pick his own routes up mountains. His years of studying rock and glacier formations as he surveyed and composed his maps had given him very sound judgement in estimating which ridges and rock faces would be most accessible. In the event, Maccarib proved to be the easiest of the climbs the men made in the valley of Portal Creek. The point at which Bridgland and Martin stopped to take their readings and photographs was not the true summit but only a very high ridge. The summit would have to wait another couple of weeks, until the weather was clearer. Despite the frequent rain and snow that day, the camera stations they occupied allowed them to see the surrounding country and to plan their next forays.[42]

On 8 and 9 July, M.P. made two successful first ascents, Chak Peak (7,460 ft [2,275 m]) and Mt Aquila (9,270 ft [2,825 m]), respectively. The first day's weather was very cloudy so the surveyors had to wait on the summit of Chak until five in the afternoon before clearing skies permitted photography. On the second day snowstorms and rain plagued the men much of the afternoon, but again they persevered and were able to make their ascent to gather the necessary readings and photographs. Back at camp, finding himself short on supplies, Bridgland sent one of the boys back to Jasper with some exposed plates, asking him to return from base camp with fresh plates and perhaps some fresh bread or vegetables. Snow came after dinner and persisted all night. Waking up to a white blanket over the tents, the meadows, and what they could see of the mountains, the party prudently decided to spend the day "indoors." In the evening, the snow turned to rain, and the rain continued for almost a full week.

On Friday, 16 July, the sixth full day of rain, a disgusted chief of party hiked out to Jasper for more supplies and correspondence. The walk took almost six hours but M.P. had Ed Hyatt and his assistants for company; they were finished with the work they had been doing

HIS YEARS OF STUDYING ROCK AND GLACIER FORMATIONS AS HE SURVEYED AND COMPOSED HIS MAPS HAD GIVEN HIM VERY SOUND JUDGEMENT IN ESTIMATING WHICH RIDGES AND ROCK FACES WOULD BE MOST ACCESSIBLE.

along Astoria River and were heading north to establish stations on the north bank of Miette River. Despite the cool weather, the wild strawberries were already appearing on the flats around Jasper and along the edges of the paths they followed into town. Just as they crossed the Miette, they took a moment to watch the work of a large colony of beavers living at the river mouth. That night, the men stayed at the Jasper camp. M.P. remained there the next day to answer correspondence and develop plates. Sunday, 18 July he spent hiking under heavy showers back to his camp at Portal Creek.

The beginning of the work week dawned clear and bright, and M.P. could finally set off with high hopes of making yet another first ascent, this time of Mt Majestic (10,125 ft [3,086 m]), but, caught in heavy clouds half-way up, he was forced to retreat. He succeeded in summitting the elusive mountain the next day but because of the heavy snow pack on the peak he took his readings and photographs from a snowfield about 150 ft (45 m) below it. With so much recent adverse weather, he was especially wary of sudden changes in the skies. He and Martin worked as quickly as they could to shoot plates and capture readings. The next day, he had the packer rig up the horses and move through Maccarib Pass on into the valley of Maccarib Creek, a branch of Meadow Creek. The pass lies about 7,100 ft (2,160 m) above sea level and leads into an open valley with scattered timber along both banks of the creek. They set up camp about five km beyond the pass on the west shoulder of Mt Clitheroe, in among the creekside trees. Hidden in the heart of the mountains, this infrequently penetrated valley seemed a veritable jewel to the surveyors. Connecting to the Tonquin Valley and the marshy meadows of the east shore of Amethyst Lakes, it featured a lush strip of forest on the gentle slopes and, higher up, wide alpine meadows inhabited by caribou. These meadows and ample stands of green spruce and balsam offered innumerable options for campsites and grazing for the horses, while the southern end of the lakes made an aesthetically wonderful backdrop to the camp, with picturesque views of rockslides and steep moraines formed when massive glaciers had retreated up the huge wall of the Ramparts.

While the packer moved camp, Bridgland and Martin occupied stations on Mt Clitheroe (9,012 ft [2,747 m]), adding it to their very impressive roster of first ascents in July. The mountain had not yet been named when they climbed it, so Bridgland suggested the name it bears today—which comes from a municipal borough of Lancashire, one of

Photograph by M.P. Bridgland of the head of Circus Valley and Mts Vertex, Majestic, and Estella from Vertex II stn, Stn 8, no 64, direction west, 1915. Courtesy Library and Archives Canada R214.

Photograph by M.P. Bridgland of Tonquin Valley, Amethyst Lakes, and the Ramparts from Mt Clitheroe, Stn 11, no 88, direction southwest, 1915. Courtesy Library and Archives Canada R214.

the hillier districts of England—apparently because of the magnificent view of Amethyst Lakes it afforded from its summit.[43] Because M.P. must have occasionally tired of photographing endless mountain scenes, it is all the more noteworthy that to him "the lakes, reflecting in their placid waters the overshadowing range with its varied snow-clad outlines and dark rugged cliffs, form[ed] a picture that is seldom equalled."[44] Luckily, he managed to make his photographs before the weather yet again broke with violence. Thunder and lightning lasted through the night while the tented men did their best to stay dry. Morris made it his special mandate to ensure that his men always kept their boots and feet from getting wet. Even in dry weather, he would insist that boots and feet be dried by the fire every evening. The transit and camera aside, a surveyor's priority for a successful summer of climbing was clean, dry, blister-free feet.

The following week's weather was consistently cloudy and dull in the mornings and relatively clear in the afternoons, with occasional thunderstorms in the evenings. Over five successive days, the men took stations on Surprise Point (7,874 ft [2,400 m]), Thunderbolt Peak (8,744 ft [2,665 m]), both in the upper Astoria River valley, Tonquin Hill (7,861 ft [2,396 m]), located on the continental divide, Mt Arris (8,875 ft [2,705 m]), and Mt Maccarib. The latter involved a second attempt, but this time they reached the true summit. They could see the Tonquin Valley and the beautifully coloured Amethyst Lakes to the west, and rising above these the magnificent escarpment named the Ramparts, which forms the continental divide as well as the boundary between the provinces of Alberta and British Columbia. La Montagne de la Grande Traverse (11,034 ft [3,363 m]), the most significant mountain in the Athabasca River valley during the period of the fur trade, towered above all surrounding peaks. South and west around behind this mountain lay the five- or six-day horse or snowshoe trip up the Whirlpool River and over Athabasca Pass to Wood River, a tributary of the great Columbia River. For four decades between 1812 and 1853 the traverse formed the principal obstacle in the transcontinental route between Hudson Bay and the Pacific Ocean. The mountain's name had informally changed to Mt Geikie and Mt Fitzhugh by Bridgland's day, and was formally changed in March 1916 to Edith Cavell in honour of the famous English wartime nurse (1865–1915) who was executed on suspicion of having aided Allied prisoners to escape. The precipitous walls and sharp pinnacle of Oldhorn Mtn (9,843 ft [3,000 m]) rose to the immediate south. To the southwest the dark rocky peaks of Mt Fraser (10,726 ft [3,270 m]) and Mt Erebus (10,233 ft [3,120 m]) protruded impressively from glistening snowfields.

On Tuesday, 27 July, Bridgland and Martin's ascent of a high ridge on Vertex Peak (9,702 ft [2,957 m]) concluded their work in the Tonquin and Portal Creek valleys. On Wednesday the crew packed themselves and the horses back to the main camp at Jasper, and happened into Hyatt on the way. His achievements during the ten days of Bridgland's first ascents were equally impressive, for the number if not the height; he had surveyed the prominent peaks on both the north and south sides of Miette River, including Mt Henry, Emigrants Mtn, Mt Elysium, Indian Ridge, Clairvaux, the Whistlers, Marmot Mtn, and Muhigan Mtn. (Of these, Indian Ridge alone exceeds 9,000 ft [2,750 m].) Clearly, the decision to split the summer's work over two parties had been an inspired one. The only disadvantage proved to be the odd overlap, as was the case when Hyatt climbed and surveyed from Marmot Mtn just after Bridgland. In all probability, the men had a good laugh over the misunderstanding, as the peak had not been a difficult one to ascend and relatively little time had been wasted by the repetition.

❋ A Day Off, A Life Lost ❋

THROUGHOUT THE VALLEYS EXPLORED IN JULY, many obstacles presented themselves: on the lower slopes, dense underbrush and windfall made climbing and hiking a struggle, while higher up, snow slopes and glaciers had proven difficult to traverse. However, a veteran would have encountered and overcome all these often enough in the past. Nothing though could prepare Bridgland for the next instance of adversity he would encounter, one of a very different type. On Thursday, 29 July, the whole crew took a day of rest and relaxation. Hyatt and presumably some of the other men decided to go for a swim in Horseshoe Lake/Lac Beauvert while Bridgland straightened up the camp and prepared for the next outing, which was to take them east along the south bank of the Athabasca. He wrote letters to his superiors informing them of his plans for the following week and went into town to post them. Either on his way back to camp or perhaps later in the evening, he was informed by one of his men that Hyatt, his colleague and young friend for the past several summers, had died.

NOTHING THOUGH COULD PREPARE BRIDGLAND FOR THE NEXT INSTANCE OF ADVERSITY HE WOULD ENCOUNTER, ONE OF A VERY DIFFERENT TYPE.

Whether in an attempt to remain professional despite the tragedy, or with a stoicism indicative of his sense of loss and shock, in his surveyor's diary Bridgland only recorded starkly that "Hyatt was drowned."[45]

On Friday, a police inspector came up to the camp, held an inquiry into the circumstances connected with the drowning, and then released the body to Bridgland's custody. A second unrelated difficulty presented itself when cook A. McKinnon quit the crew because of an injured knee. With no choice but to leave the remaining three men to continue the survey as best they could, Bridgland boarded the train on Saturday to have Hyatt's body embalmed at Edmonton. Thereafter, he continued south and then west by train, reaching Revelstoke, British Columbia, Hyatt's home town, on Monday, 2 August. (One could not make the roundabout Jasper-Edmonton-Revelstoke trip much more quickly today.) There, he helped with the arrangements for the funeral and burial of his young assistant, dead at twenty-three.

"HYATT WAS
DROWNED."

Although it was probably the last thing he wanted to do, Bridgland had to make arrangements to replace Hyatt on his team. He returned to Calgary late on Tuesday and spent until Saturday working at his office in the Thomas Block, developing plates taken in late June and early July, and contacting his superiors so that they could send out another assistant surveyor. On Sunday, 8 August, he took the train to Edmonton where he met J. Robertson whom he employed to replace McKinnon as cook. Surveyor General Deville had promised to send out a qualified assistant as soon as one could be found. Of course, owing to the war, few men were available. Bridgland and Robertson continued on and arrived in Jasper the same night. There, he promptly caught up on the men's activities during his absence and related the sad details of Hyatt's funeral and his family's grief. The team had moved east along the Athabasca and set up camp on the Maligne River. The weather had been too stormy for photographic work on most days, but "the boys" had occupied a few stations along the Maligne and had climbed a peak far enough up the road to Maligne Lake that it afforded a view of Medicine Lake. Bridgland moved their camp downriver to the Maligne's confluence with the Athabasca.

The pace of work and the organization of the survey party suffered with the death of Hyatt. Norfolk, who like Hyatt had been with Bridgland for at least two seasons, had expressed a sincere interest in the phototopographic work, and he knew how to handle and operate the instruments. Bridgland temporarily promoted him to assistant and increased his salary from that of labourer. The Department of the Interior sent out another assistant, but by the time William Cuthbertson (b. 1891) arrived from Perth, Ontario, three weeks later with no mountaineering or phototopographic experience, Bridgland had Norfolk continue on as the assistant surveyor; he had proven his ability with the camera and transit, and had been able to produce views that even Bridgland's exacting eye considered satisfactory. This arrangement left Cuthbertson the transit work, in which he was already trained and practised, and Norfolk helped him by pointing out the stations the party had occupied earlier that summer. Once they had spent some days assessing their respective skills, these two men travelled together, and Bridgland carried on with the other, less experienced men.

✳ *August Along the Athabasca* ✳

FOR THE REMAINDER OF THE MONTH OF AUGUST, Bridgland led the entire party along the south-and-east bank of the Athabasca River, making surveys as they proceeded of the valleys of the lower Maligne, Rocky, and Fiddle rivers. For three weeks, he and his men worked in sweltering heat from Jasper to Pocahontas. The Athabasca is very broad in this stretch, widening into shallow Jasper Lake and then Brûlé Lake before narrowing back into a river as it leaves what was then the park's eastern gate (Black Cat Mountain and Brûlé Lake have sat outside the park since its border was redrawn in 1930).[46] Several mountain ranges terminate along both sides of the valley, giving the landscape a more broken appearance than in the valleys like that of the Columbia River, where the river runs between entire parallel ranges, the Selkirks and Monashees. The land itself was marshy, the riverside trail often washed away by storms or submerged by high waters from the wet summer run-off. Trails in the higher valleys were generally easier to follow.

When he reached the active hamlet of Pocahontas, Bridgland sent a letter in reply to one from Surveyor General Deville enquiring about names for features in the Crowsnest Forest Reserve mapped the previous summer. In a touching moment, he suggested that one of the names, for a Mt Patricia, be changed to honour his deceased assistant. The reason for a mountain named for him was simple: "It was climbed by Mr. Hyatt in 1914."[47] M.P. did not urge the change, only suggested it modestly, but it would have meant a great deal to him to have it accepted; he must have

Photograph by M.P. Bridgland of the view of Athabasca River, Pocahontas (*centre left*), Roche Miette, and rail lines on either side of the river, from Bedson Ridge, Stn 75, no 620, direction southeast, 1915. Courtesy Library and Archives Canada R214.

been missing him keenly by this point in the season's work. In the event, while no Mt Patricia appears to have survived, neither is there a Mt Hyatt; apparently Deville did not adopt Bridgland's suggestion.

The men worked out of Pocahontas for several days before trekking up the marked trail to Miette Hot Springs, watched over by a large herd of sheep that seems eternally to occupy the slopes of Roche Miette. On the flats just north of Fiddle River, saskatoon berries were abundant, and along the trail to the springs the men helped themselves to a fine crop of raspberries. The rapid rise and fall of mountain streams was not uncommon; as they hiked along the narrow valley of the Fiddle, M.P. told his men a story he had heard about the river's caprices: "An early store-keeper who was crossing Fiddle river with a team of horses and a wagon load of supplies, stopped in the stream to remove some of the boulders and while doing so heard a roar above him. So rapidly did the water rise, that to save himself he had to abandon both team and wagon."[48]

Photograph by M.P. Bridgland of Ashlar Ridge (*foreground*), Fiddle Range (*middle ground*), Roche à Perdrix (*background*) from Roche Miette, Stn 83, no 680, direction eastnortheast, 1915. The shape of a large bass fiddle, lying on its side, can be traced in the outline of the Fiddle Range; hence its name. Courtesy Library and Archives Canada R214.

The hot springs, located on the west branch of Fiddle about nineteen km (twelve mi) from Pocahontas, ranged in temperature from 100°F to 130°F (38°C to 54°C). They were no more developed as a tourist attraction than they had been four years earlier, when DLS Herriott had been sent to survey for an entire town at the river mouth and a luxury hotel up at the springs. Bridgland allowed his men the chance to test the warm rocky pools personally and gain some relief from the black flies. Four years later, in 1919, the Pocahontas coal miners went on strike, and with no work to do in the daytime they headed for the hills and built Miette Hot Springs' first log hot pool. It would be another decade before the springs underwent further development.

On Sunday the men moved their camp up the Fiddle River Canyon, a fair distance beyond the springs and above the torment of August

heat and "bulldog" black flies. The little meadow covered with a luxuriant verdure seemed a veritable haven; the water was clear and the
weather cool. When Bridgland established his camera stations at the
top of the mountain overlooking their campsite, he named it Fiddle
Creek West in his transit notebook, but added the word *Utopia* in the
upper corner of the page. At his later suggestion, the peak was assigned
the name it retains today: Mt Utopia (8,537 ft [2,602 m]).[49]

On Tuesday, 17 August, the smoke from area forest fires grew thick
enough to halt topographic work. Climbing was not only hazardous
when the visibility was so poor but also pointless as far as the visible
detail in photographs of the landscape was concerned. That evening
Cuthbertson arrived with a shower of rain that continued for two days.
When it let up, the haze from the fires returned and the survey party
could hardly believe its continued bad luck. Alternating daily between
rain and hot, dry, smoke-filled air, weather precluded surveying for ten
full days. Bridgland still put his crews to good use but they can't have
been pleased with his decision to send them down on the floor of the
Athabasca valley to work in the heat on a traverse of the GTPR from
the park's East Gate to Jasper. Bridgland planned to use it as a base for
a triangulation survey that would confirm his phototopographical
survey. The object of a traverse, or secondary triangulation, is the determination of the exact location of camera stations relative to a known
longitude and latitude. He searched out the DLS posts that the earlier
surveyors of base lines and meridians (St Cyr, Hawkins, and J.B.
McFarlane) had staked because the longitude and latitude of these had
already been determined and verified. Locating from each of these at
least two of the cairns built at his camera stations, he used the transit
to measure the angles between these points. He would also take an
azimuth reading of at least one well-defined point at each post. He
plotted all this information on a rough draft of his triangulation.

On the last day of August, Bridgland sent two men with the horses
to a ferry north of Pocahontas where they would be taken across to the
west bank of the swift Athabasca. The rest of the party hoped to cross
the river from Pocahontas to Bedson, home to Miette, the CNoR
station, which as Bridgland had noted on his way into Jasper two
months earlier, corresponded to the GTPR station across the river at
Pocahontas. At Bedson they set up their camp for September.

ALTERNATING DAILY
BETWEEN RAIN
AND HOT, DRY,
SMOKE-FILLED AIR,
WEATHER PRECLUDED
SURVEYING FOR TEN
FULL DAYS.

Photograph by M.P. Bridgland of the view of Athabasca River valley's Jasper Lake from Mt Esplanade, Stn 62, no 507, direction eastnortheast, 1915. As the party worked back to Jasper, the lower light of late September yielded photos with less clear resolution. Courtesy Jasper National Park.

✳ Autumn in the Valley ✳

FROM BEDSON the crew spent all month working back towards Jasper along the north and west side of the valley, surveying several peaks along each of the valleys of the major tributaries they crossed, namely Moosehorn Creek and Snake Indian and Snaring rivers. They found trails up the first two valleys, but none up the Snaring. (No surprise there: veteran backcountry travellers consider the canyon up the Snaring as the most inaccessible section of Jasper National Park.) The trail along this bank of the Athabasca was not as prone to flooding but the surveyors did encounter one section that had been dammed by beaver near Coronach Creek, downstream from Snake Indian River. It was so flooded that the horses had to swim across while the men bushwhacked their way to a higher point to find a ford. The weather was somewhat kinder in September than in August but even when it was fine the time for optimal photography was growing shorter by the day. The late autumn sun rose lower in the sky, and the shadows in the valleys, creeping higher up the mountainside, seemed to intensify in their darkness, making good black and white photography quite a challenge.

At the end of September a small party headed up by Cuthbertson and Norfolk crossed the Athabasca to Jacques Creek. They pitched camp near Interlaken, a stop on the GTPR line. From there, they extended their work east, into the upper valley of Rocky River. The remainder of the party occupied two stations on the Maligne Range. Inclement weather rendered these surveys less successful than Bridgland had hoped. Once the full party returned to Jasper in early October, everyone filled in the traverse information for the triangulation base along the GTPR line from Jasper to the Yellowhead Pass, completing the work on 13 October within sight of the mountain that would soon be renamed in honour of the nurse who had been executed by a German firing squad the day before. M.P. made arrangements for storing the outfit and for leaving the horses with the Otto Brothers outfitters until such time as they were again needed by a surveyor or other Department of Interior employee. On Friday, 15 October, they carefully packed up the two transits and cameras and sent them to the Ottawa survey office for cleaning, calibration, and winter storage. They then took trains to Edmonton and on to Calgary where M.P. paid and discharged his boys.

THE WEATHER WAS
SOMEWHAT KINDER IN
SEPTEMBER THAN IN
AUGUST BUT EVEN
WHEN IT WAS FINE THE
TIME FOR OPTIMAL
PHOTOGRAPHY WAS
GROWING SHORTER BY
THE DAY.

✳ Once Home in Calgary, the Challenge of Map-making ✳

ABOUT A WEEK PRIOR TO LEAVING JASPER Bridgland had written to Surveyor General Deville seeking clarification of his and others' employment for the winter. Hyatt had been his primary assistant in the detailed work required in making his maps, and Bridgland was quite concerned about having such a large map to make without qualified assistance. He proposed to offer a job for the winter to Martin, who had sufficient training to assist in plotting the survey. The department would not agree to this arrangement so Bridgland wrote again, arguing that, as Martin was already familiar with the work and was qualified, he would make the most economically efficient candidate to render much-needed assistance. Cuthbertson, who had been engaged by the department in

Ottawa but who had no practice in phototopographic plotting, was already assigned to the project, but Bridgland persevered in his argument that his work would be seriously hampered if his assistant required constant instruction. Eventually permission arrived to hire Martin.

Although returning home was a pleasure for M.P., who was greeted by his wife and seven-year-old son, Charles, and although the summer had been one of hard physical labour, a different challenge awaited him in Calgary. Transforming over seven hundred panoramic photographs into a detailed topographic map without the aid of any computer seems a Herculean task today. At the start of the summer, M.P. had at his disposal only very sketchy maps by which to reconnoitre the valleys and meadows, the mountains and passes. These held practically no information about the relative positions of the ranges, or the height, breadth, or pitch of individual peaks. Most of the information on them, gathered by the railway surveyors, consisted of rough sketches of rivers bridged by the railways. While in the field he had simple "sectional sheets" which showed the principal meridians and had St Cyr's, Hawkins's, McFarlane's, and Matheson's straight township grids drawn over the areas directly adjacent to the rivers. As he surveyed during the summer, onto these sectional sheets he would pencil hatch marks to outline the contours and general positions of the mountains, but these marks served chiefly as an *aide-mémoire* of the mountains he had surveyed, not as a record from which to compile precise topographical information.

On 20 October, a set of 10" x 14" (25.4 x 35.5 cm) bromide enlargements arrived in Calgary from Ottawa to which, periodically during the summer, M.P. had sent batches of the glass plate negatives for development. It was, as mentioned earlier, Deville's policy that all developing and printing should be done in one central laboratory so that he could ensure consistency of quality, format, and size; the Department of the Interior had invested heavily in an enlarging camera made specifically for the purpose. Because the photographs would constitute the main reference for the drawing of maps, with the transit notes and the triangulation calculations from the base-line traverse complementing them, it was of the utmost importance that they be developed as meticulously as Bridgland had produced the initial negatives.

What a hash! The first batch of prints was without exception very poor: sharp definition and contrast were lacking. M.P.'s heart must have sunk. And then risen to a simmer. He immediately wrote to inquire into their poor quality but received a reply stating that the printer

TRANSFORMING OVER SEVEN HUNDRED PANORAMIC PHOTOGRAPHS INTO A DETAILED TOPOGRAPHIC MAP WITHOUT THE AID OF ANY COMPUTER SEEMS A HERCULEAN TASK TODAY.

could achieve no better results. Comparing these enlargements with those from the previous two years' survey farther south, he determined that the poor quality of the Jasper prints resulted from bad darkroom technique in the printing process. He relayed this information to the department, insisting that the maps' accuracy depended on clear prints. After sending samples from his previous years' work to buttress his case, he was promised by Deville that he would be sent satisfactory prints. About to receive by Royal warrant the distinction of a Companion of the Imperial Service Order in recognition of "the sterling services he ha[d] rendered to Canada and to the Empire by his researches," Deville had remained as keen as Bridgland had grown about the phototopographical method. Even five years later, the innovation was still being widely hailed; in 1921, a copy of Deville's camera was used in the mapping of the north slope of Mount Everest.[50] Deville shared and supported—at this point in his distinguished thirty-year career as surveyor general, he had come to rely on—Bridgland's unyieldingly high standards.

While waiting for the photographs, M.P. started plotting a new copy of the triangulation he had drafted in the field. When an improved set of enlargements arrived he embarked on the most tedious and laborious of the mapping tasks: transforming all his photographic information into contour lines, a process called iconometry. This work was very hard on the eyes, highly stressful, and unappealingly exacting. "The office work" of phototopographic surveying, Wheeler wrote in 1920, "occupies at least twice the time of the work in the field."[51] Not surprisingly, staff shortages in the Calgary office seemed to occur when it was time for this work, so Bridgland often had to work overtime. His method was to select a photograph from his summer's work and pin it to the drawing board in front of him. The board was covered with strong paper ruled with lines marking the principal point, horizon line, and principal line in his field of vision, as well as the limits of the focal length and other lines that helped determine the angles in a photograph's landforms. Because of the notches he had carved into the metal backing of the camera, each print had small darts of exposed image in the otherwise white border; these marked the horizon and principal lines in each view and also ensured that, even if there were slight variations in the dimensions of the enlarged prints, the field of view could always be accurately calculated. He lined up these notches with the lines on the board and then was ready to begin his calculations.

"THE OFFICE WORK" OF
PHOTOTOPOGRAPHIC
SURVEYING, WHEELER
WROTE IN 1920,
"OCCUPIES AT LEAST
TWICE THE TIME OF
THE WORK IN THE
FIELD."

Photograph by M.P. Bridgland of grid applied to Bridgland's photograph of the Amethyst Lakes and the Ramparts from Mt Clitheroe, Stn 11, no 88, direction southwest, 1915. From M.P. Bridgland, *Photographic Surveying*, 44.

✳ *Mapping from the Phototopographic Surveys* ✳

HE BEGAN BY PLOTTING the most prominent points in each photograph, and then drew individual contour lines to show the literal shape of the mountains, where the pitch was steep and where it was gradual, where rock faces dropped off in sheer cliffs, and where meadows gave way to muskeg and marsh. Taking two views from different stations showing the same country, he ruled onto the prints the principal horizon lines (their positions were marked by the notches in the negative's border) and the picture plane (essentially a slanted grid that covered the foreground and gave it a sense of depth). He then identified and numbered on each of the pair of photographs as many common landmarks as he could. By determining their relative positions through a detailed series of steps and calculations, he then plotted these points on the plan.

In the field, M.P. had levelled his camera so that he had a consistent horizon point in each of his several hundred photographs. Using his transit, he subsequently measured the precise orientation, or horizontal angles, of each of these views relative to the survey station. Back in the office and out of the uncooperative weather of the summer of 1915, he could determine the relative positions of the mountain peaks by first plotting the camera station on paper, then measuring the distance of the identified points from the principal line. He would trace this line onto a slip of paper, using a separate slip for each view. Once he had traced two slips showing the lines between identical mountain peaks from different camera stations, he would place one slip over the other. The point at which the lines intersected could thus be marked on the plan as the true position of the mountain in question relative to the camera station. He marked and numbered each point as he plotted it, slowly filling his map plan with small numbers that identified mountain peaks or other significant ridges. In order to save time and to determine the intersections of lines, he often inserted needles at each point; between two points he would string fine silk threads or hairs, instead of drawing lines on the plan from each station.

With a plan that now delineated the rivers' courses and showed the position of each mountain's summit by means of a small dot, he next faced the challenge of determining the elevation of each of the identified points, except for the camera stations and more prominent peaks, the elevations of which were read in the field with the transit. For this task, he used an instrument for computing elevations, a tool originally devised by Donaldson B. Dowling and Hugh Matheson of the DLS. This unnamed device was to plotting what the slide rule was to mathematics; it bore some physical resemblances, as well. It consisted of two arms of brass fastened rigidly together in a right angle with two sliding bars that could move on the horizontal arm, and one arm that revolved around a centre point on the instrument. The moving parts were made of clear celluloid and each of the bars was graded. Although the tool was highly sophisticated for its time and had to be aligned according to very specific parameters (the centre had to be positioned directly over the camera station, one of the sliding arms was aligned with the principal line of the trace of the view, and so forth), it worked essentially as a series of rulers that measured relative elevations.[52]

Using the plotted points of elevation and the photographs as a guide, Morris could then draw his contour lines on the tracing tissue, trying to follow the irregularities of the surface as they appeared in the photographs. Aside from having the photographs constantly before him, his field sketches and his knowledge of geology born from climbing experience served him in this exacting exercise. When photographs failed to give him the detail he required, he relied for the shape of features on his knowledge of how certain rock types tended to bend or break as they formed themselves into mountains. Meanwhile, the marshes, lakes, and rivers with relatively small fluctuations in elevation were plotted by means of the picture plane, or, as he termed it, the perspectometer. This series of squares with its distance line equal to the focal length of the photograph allowed him to draw in the outlines of flat features, square by unrelenting square. Gradually, on a 1:62,500 scale (just over 1 inch per mile) and 100-ft contour intervals, he produced his monumental *Map of the Central Part of Jasper Park, Alberta* in six sheets— northwest, northeast, west central, east central, southwest, and south-east—and sent it to Ottawa for final production by the surveyor general's office and publication by the Department of the Interior.

The map of the 1915 Jasper survey proved the rule that mapping took longer than photographing; Bridgland and his office staff worked well into the following spring to complete it. Meanwhile, after the long process of rendering the topographic details came the marking of official names for each of the prominent peaks. For each mountain he surveyed that did not already have a name, M.P. was asked by the surveyor general to suggest one. As a letter from him to Bridgland dated 6 December 1916 clarifies, Deville had very practical reasons for preferring some names over others: "Amethyst lakes, Mt Blackhorn, Mt Erebus, Roche Noire, the Forum, the Rostrum, Clairvaux creek, Portal Peak and Mt Cairngorm are good names. Meadow creek and Mt Aquila may do. I am not favourably impressed with names like Longview or Clearview; there is nothing distinctive in a wide view being obtained from the top of a mountain."[53] Deville also had an exceptional knowledge of Jasper history and often suggested names that would honour explorers or commemorate historic events. In some cases, he would suggest the reinstatement of names once given by early explorers or fur traders to a particular land-mark but that had fallen out of use. Although he consulted no Natives to learn their names for mountains or other features, occasionally he retained or selected names from Native languages. The Athabasca River kept its name, as did Mt Tekarra, named by James

THE MAP OF THE
1915 JASPER SURVEY
PROVED THE RULE
THAT MAPPING
TOOK LONGER THAN
PHOTOGRAPHING;
BRIDGLAND AND HIS
OFFICE STAFF WORKED
WELL INTO THE
FOLLOWING SPRING
TO COMPLETE IT.

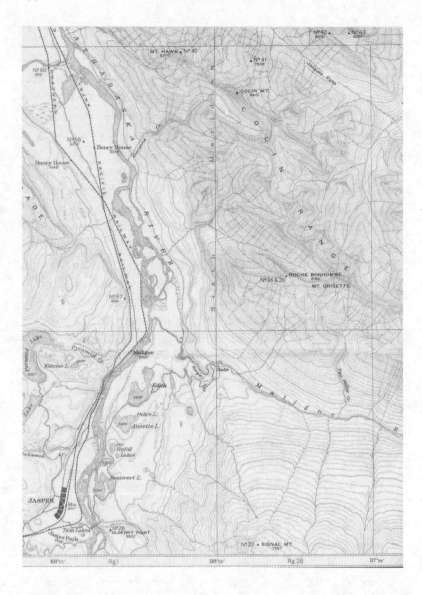

Map of the Central Part of Jasper Park, Alberta. From Photographic Surveys by M.P. Bridgland and A.E. Hyatt, 1915. 1:62,500. Ottawa: Dept. of the Interior, [1917] (detail). The Athabasca River valley at and down river from the town site of Jasper is depicted in this detail. Courtesy William C. Wonders Map Collection, University of Alberta.

Hector in 1859 for his mixed-blood Iroquois guide. Bridgland named Chetamon Mtn, using the Stoney word for squirrel because two rocks on the mountain's arête resemble its shape; Mt Maccarib, after the

Quinnipiac name for caribou, which he saw near the peak; and Kataka Mtn, Muhigan Mtn, Makwa Ridge, and Wabasso Lake after Native words for fort, wolf, bear (or perhaps loon), and rabbit, respectively.

✳ *The Art of Bridgland's Photographs* ✳

THE STRESS ON KNOWING rather than imagining runs deep in the Canadian mind. It is a necessary illusion by which Canadians feel they can exert a needed sense of comprehensive control over the nation's huge land mass. Of course, we cannot, but the illusion of our being able to do so is one that we cherish because it keeps at bay the knowledge of how few of us there are for such a vast land mass. Bridgland's photographic work, by comprising a portion of an enormous Canadian project that gave way in subsequent decades to aerial photography of the entire country, can most clearly be understood in the inventory tradition. However, a photograph, regardless of how systematically one takes it or how primarily it figures into a larger project, always involves a measure of interpretation. Schooled as an empirical surveyor, M.P. would doubtless have bristled at such a suggestion; indeed, his temperament suggests very clearly that he did all he could to systematize his work in an effort to leave his personal touch out of it. (That would come later, in retirement, when he made selections of his personal photographs and hand-coloured them as illustrations for his talks.) The decree that all surveyors' photographs be sent to Ottawa for developing under a common procedure at the same lab by the same technicians shows how clearly his approach to his work aligned with the purposes of the federal inventory by the Dominion Lands Survey.

THERE IS NO SIMILAR BODY OF WORK ANYWHERE THAT CAN COMPARE WITH ITS COMPREHENSIVENESS, ITS UNDERSTANDING OF MOUNTAINS, THEIR QUALITIES, THEIR FOIBLES, THEIR VARIOUS MOODS AND DEMEANOURS.

However, any time spent with Bridgland's surveys, whether photographs or subsequent maps, leaves one deeply impressed with the artistic dimension of the man's work. Painstakingly systematic though the fruits of it evidently are, it is also individualized. There is no similar body of work anywhere that can compare with its comprehensiveness, its understanding of mountains, their qualities, their foibles, their various moods and demeanours. And there is art in this inventory, for photographs necessarily *construct*, they do not just record, their subject matter. This construction occurs through various practices involving the capacities and constraints of the technology itself and the ways in which it does and does not encourage us to think.[54] All photographs, like all maps, conflate art and science, experience and interpretation, documentation and picture, reality and construction. Every photograph lies somewhere on the continuum between the entirely objective and the wholly impersonal. Concurring with Paul Berger, Rocky Mountains Repeat Photography Project (RMRPP) researcher Jenaya Webb notes that the viewer of a photograph "'locates' [it] along this continuum[;] that is, how he understands it, is determined by more than just the photograph itself."[55]

HE SHOT
PHOTOGRAPHS
IN HIS PARTICULAR
WAY, ONE THAT
HIS EYE JUDGED
PRODUCED THE
MOST SATISFACTORY
RESULTS.

This idea of the role of art, as it bears even on systematic survey photographs, might diverge from what we have in mind about art when viewing, say, the celebrated landscape painting of Tom Thomson (1877–1917), one year Bridgland's senior and at the height of his brief career in Bridgland's Jasper field season. But consider the element of the personal in Bridgland's choice of stations from which to take a panorama of photographs, as well as in the use of filters to diminish the effects of shadow for the purposes of revealing the diverse qualities of mountainscapes, or in his inclination to use stops as small as f/32. He shot photographs in his particular way, one that his eye judged produced the most satisfactory results. Consider even his remarks about light:

> Views looking under the sun will, in bright light, require two or three times the normal exposure. Light is much stronger in the middle of the day than in the early morning or late afternoon, and stronger also in May or June than later in the season. In Canada, after about the first of October, as the sun becomes lower, the shadows in deep mountain valleys become more intense, and it is impossible to obtain good views except near noon and under very favourable conditions. At high latitudes the light is more actinic [susceptible to chemical change by the violet and ultra-violet wavelengths of the spectrum] than at sea-level, and also varies according to the latitude of the place.[56]

These particular factors particularly influenced the way that Bridgland conducted his work and oversaw that of his assistants. Clearly, much interpretation is necessary in the making of a photograph that aids the drawing of an accurate map. In the same way as this statement helps us see how the photographer needed to adapt his work to light conditions, so we can appreciate that the photographs provide a version of landscapes that of necessity adopted a much higher prospect point than one sees in most paintings of landscape by Thomson or his successors in the Group of Seven. But just as Thomson was famous for his sense of urgency to get a painting down on board or canvas while the light was just right, we can sense a kindred spirit in Bridgland, who, if weather granted him the choice, made his climbs and his photographs in order to obtain the light he considered best for the photographs that he had in mind.

Bridgland's photographs are far from flat. Pre-eminently, they reveal relief in the terrain. Perhaps ours would as well if we were making them in order to meet the subsequent need to produce a two-dimensional topographic map from them. Still, one can readily appreciate how, over the course of more than a quarter-century, he allowed his own understandings of photography to colour his work, his own love for mountains to influence his photographic renderings of them. Can one tell a survey photograph made by Bridgland from one made by a colleague? Yes and no: all phototopographic work shares common attributes but distinctive qualities, such as sharpness or degree of overlap or minimization of the effects of light are apparent to the trained eye. Most of us can recognize a painting by a member of the Group of Seven, a school of painters. But few of us can assuredly identify a Carmichael as distinct from a Jackson every time. Meanwhile, probably very few of us have stopped to consider that a Group of Seven artist could have been inspired by a DLS phototopographer. But consider one remark about how Tom Thomson composed:

THOMSON'S ACTUAL
SKETCHING TECHNIQUE,
WHICH HE SHARED
WITH MOST OF THE
PAINTERS WHO LATER
BECAME THE GROUP
OF SEVEN, AT TIMES
APPEARS TO EMULATE
THE ACTIVITY OF
THE CAMERA.

> Thomson's actual sketching technique, which he shared with most of the painters who later became the Group of Seven, at times appears to emulate the activity of the camera. He strove for the emphatic reality of the moment. The consequent note of urgency, although evident in the finished work, is, in his better sketches, saved from appearing sloppy by the intense rigour of their reality. The open-sided horizontal composition so prevalent in Group of Seven sketches is also a normal characteristic of the snapshot.

Dennis Reid advanced this view in an article written after a collection of forty of Thomson's photographs surfaced and were published in their entirety for the first time in 1970.[57] The idea that our country's surveyors and painters shared perspectives on landscape is not a topic that *Mapper of Mountains* can do more than broach, but its very suggestion contributes another way of seeing Bridgland's enormously valuable work.

✳ *Bridgland's Monumental Achievement* ✳

WITH THE MAP COMPLETE and the numbers tallied for a cost assessment of the survey, it was discovered that the work produced in the summer of 1915 broke all records. On average, a topographical surveyor could expect to collect in one season enough data to map between 400 mi^2 and 500 mi^2 (1,000 km^2 and 1,300 km^2). Bridgland doubled this average, mapping about 950 mi^2 (2,500 km^2). This feat is especially impressive when we consider all the factors working against the surveyors, as Bridgland noted in the report he submitted to Deville:

> The season of 1915 was very unfavourable for photographic work. From June 24 to October 15, there were only forty-eight fine days out of a total of one hundred and fourteen [or 42 per cent] and even on the fine days there was nearly always a slight haze or heavy cloud shadows. In July, September, and October, much delay was caused by wet weather while in August twelve days were lost owing to smoke. These unfavourable conditions, combined with the unusual width of the valleys, made it very difficult to obtain satisfactory views.[58]

After all the expenses for field and office work were tallied, Bridgland found that the survey had cost an average of only $4.20 per mi^2 ($1.60 per km^2), a savings of between three and four dollars per square mile compared to the surveys completed by his mentor Wheeler in the Main Range of the Rocky Mountains and in the Selkirks, two ranges west of the Main Range, and by the survey of the Main Range in 1892 by the redoubtable J.J. McArthur, DLS. M.P.'s efficiency did not go unnoticed: just a few months later, the surveyor general remarked in a letter to the director of Forestry at the Department of the Interior that Bridgland was the department's best topographic surveyor.[59]

Bridgland's strict work ethic, physical stamina, and dedication were instrumental in accomplishing so much work in so little time, although the relatively low elevations of the peaks of the main valleys would also have facilitated quick work. In total, he and his crew had sixty-five actual working days, twenty of which involved traverse work, and forty of which were devoted to topographical work. He and his parties needed fewer than fifty days to complete the survey. This feat is all the more impressive when we recall that, at least according to surviving records, nine of the mountains that Bridgland climbed that summer had never before been ascended. One can sense, however, that it had been a long season's work. In a more retrospective moment than one usually gains from Bridgland, he wrote generally of the mountain surveyor's work with a tinge of *ennui*:

ONE CAN SENSE, HOWEVER, THAT IT HAD BEEN A LONG SEASON'S WORK.

> The first part of the day's work is the ascent to the station to be occupied, during which the party may encounter all the difficulties and experience all the *delights* so well known to mountaineers and to those who, like topographical surveyors, *only* climb mountains. At the beginning of the season the delights are paramount but towards the end the troubles assume larger proportions. It is as possible to become satiated with mountains as with anything else.[60]

If he felt satiated and, having lost Hyatt, lonely at the end of the 1915 season, his appetite for fieldwork would return after he faced a new challenge: stay in the office for a summer and write a book and make a map for tourists. ✳

✳ *Notes* ✳

1 Bridgland and his teams shot more than the final 735 photographs. Dozens of test shots needed to be made, but 735 is the number of photographs collected in photo albums, and these formed the basis for his subsequent topographic maps.

2 GTPR service was provided first in 1911 and had reached Tête Jaune Cache by the end of 1912, but survey crews had been at work in the Athabasca River valley since 1906 (Great Plains Research Consultants, "Jasper," 20, 21). The GTPR first ran trains to its terminus at Prince Rupert in April 1914. Not until October 1914 did its competitor, CNoR, which had reached Edmonton four years before the GTPR, open for service as far as Lucerne, its divisional point, which lay just across the Alberta-British Columbia border in the Yellowhead Pass. A year later it reached tidewater, too.

3 *Statutes of Canada*, 1-2 George V, chapt. 10.

4 W.F. Lothian, *A Brief History of Canada's National Parks* (Ottawa: Minister of the Environment, Minister of Supply and Services Canada, 1987), 54. The official name changed to Jasper National Park in 1930 when Ottawa passed the *National Parks Act*, but many sources continued to call it by its former name.

5 John W. McLaggan, Acting Superintendent, Jasper Park, to Robert H. Campbell, Superintendent of Forestry, Dept. of the Interior, 2 Mar. 1910, LAC, RG 39, vol. 265, file 39578.

6 For colourful character sketches of Swift in 1910 and 1911, see F.A. Talbot, *The New Garden of Canada: By Pack-Horse and Canoe through Undeveloped New British Columbia* (London: Cassells, 1911), 84–90; and Stanley Washburn, *Trails, Trappers, and Tender-feet in the new Empire of Western Canada* (London: A. Melrose, 1912), 177–78, 190–98. In Bridgland's *Description of & Guide to Jasper Park*, Swift is said to have begun homesteading in the valley in September 1891, although not all sources agree (the year 1892 is given in Great Plains Research Consultants, "Jasper," 51). By the time of the publication of Bridgland's book, the old timer had moved into Jasper (69). He died in 1940.

7 Great Plains Research Consultants, "Jasper," 55.

8 In accord with the *Dominion Forest Reserves Act* of 1906, the 14 September 1907 Order-in-Council called for the federal government to set aside 2,590 km² as the Jasper Forest Park "for the preservation of forest trees on the crests and slopes of the Rocky Mountains, and for the proper maintenance throughout the year of the volume of water in the rivers and streams that have their source in the mountains" (Canada, *Dominion Lands Act, 1873. With the Amendments and Additions Thereto* [Ottawa: Queen's Printer, 1884], 28). When it became a Dominion Park in 1911, it was reduced to a puny 2,590 km² (Lothian, *Brief History*, 54; citing federal Order in Council 1911–1338, 8 June 1911). By 1914, the protests of Commissioner of Dominion Parks Harkin, who had called the park's new size "a joke so far as utility for game protection is concerned," were heard, and the area of the park was expanded, to 13,000 km², somewhat larger than the present size (10,878 km²) of Jasper National Park. At present, the park encompasses no land east of the Front Range of mountains. The Front Range includes Folding Mountain, visible on the left of the Yellowhead Highway by motorists as they enter the park at the East Gate, and Black Cat Mountain and Brûlé Hill, visible on the right.

9 Great Plains Research Consultants, "Jasper," 29.

10 J.G. MacGregor, *Overland by the Yellowhead* (Saskatoon: Western Producer Prairie Books, 1974), 235. For two years, both the GTPR and CNoR operated trains through Yellowhead Pass, but the traffic hardly warranted two lines. In the event, amalgamation began as early as 1916, when steel began to be torn up and shipped east to meet the needs of allied forces in the Great War. The GTPR declared bankruptcy in 1919 and Canadian National Railways (CNR) took over its operations on 12 July 1920 (Cyndi Smith, *Jasper Park Lodge: In the Heart of the Canadian Rockies* [Canmore: Coyote Books, 1985], 83n); however, the CNoR

followed suit three years later. By 1921, many smaller railways had succumbed to or verged on bankruptcy. The federal government forced an amalgamation of them all under the new name, Canadian National Railways. Most of the train line through Jasper National Park today is that of the CNoR, while at many points the roadbed for the highway is the former GTPR rail bed. The CNoR's stations west of Hinton were Prairie Creek, New Entrance, Solomon, Brule/Errington, Bedson, Miette, and Jasper, while the GTPR's were Prairie Creek, Old Entrance, Park Gate, Pocahontas, Hawse, Interlaken, Henry House, and Jasper. The most concise description of the routes taken by the two lines is found in Bridgland's *Description of & Guide to Jasper Park*:

> The railways enter the Park running in a southerly direction on opposite sides of the Athabaska river, the Canadian Northern on the left [as one faces downriver] or west bank, and the Grand Trunk Pacific on the right bank. They continue up the river in this position until above Jasper lake, where the Grand Trunk Pacific crosses to the same side of the river as the Canadian Northern [about where the only crossing of the Athabasca by the Yellowhead Hwy occurs today], and at the Snaring river the two railways are only about thirty rods [165 yds (150 m)] apart. Two miles above this, the Canadian Northern crosses under the Grand Trunk Pacific, and from there to Jasper continues closer to the river and considerably lower than the latter. Beyond Jasper both railways turn sharply to the west up the narrow valley of the Miette river, coming very close together for about two miles, when the Canadian Northern crosses the Miette on a long wooden trestle. From here, both roadbeds are cut out along the steep cliffs which form the sides of the narrow valley, until near Geikie station, when the Grand Trunk Pacific crosses to the south of the Miette river, and for three miles the two tracks are laid along the same right of way. The Grand Trunk Pacific then re-crosses to the north side of the Miette, and from there on, the railways remain on opposite sides of the valley. (27)

11 *The Globe* (Toronto) 27 Oct. 1909. 8; D.J. Benham, "Jasper Park in the Canadian Rockies: Canada's New National Playground," *The Globe*, Saturday Magazine section, 15 Jan. 1910, 4, 9.

12 Sir Arthur Conan Doyle, *Memories and Adventures* (London: Hodder and Stoughton, 1924). "Western Wanderings" appeared in book form in 1994 as *Western Wanderings*, ed. and introd. Christopher Roden and Barbara Roden (Penyffordd, Eng.: Arthur Conan Doyle Society, 1994). Details of the serial publication of "Western Wanderings" are given in Richard Lancelyn Green, "Biographical Note," in Doyle, *Western Wanderings*, ed. Roden and Roden, 8, and are included in the list of Sources, below.

13 "Sir A. Conan Doyle Sails for Canada," *The Globe* 21 May 1914, 1. See also
    Robert McFetridge, "The Alberta Connection: Conan Doyle in Alberta—1914"
    http://bakerstreetdozen.com/mcfetridge.html; and http://www.ash-tree.bc.ca/
    acdschron.htm

14 *A Sprig of Mountain Heather: Being a Story of the Heather and some Facts about the Mountain
    Playgrounds* (Ottawa: Dept. of the Interior, 1914).

15 "Where the Heather Blooms in Canada. Novel Means of Advertising the
    Dominion's National Parks," *The Globe* 14 Aug. 1914, 6. The full title of the rather
    remarkably adorned booklet is *A Sprig of Mountain Heather; Being a Story of the Heather
    and some Facts about the Mountain Playgrounds of the Dominion*. Above the framed and
    recessed sprigs of heather (rose-coloured flowers, green needles) on the front
    cover, the embossed title reads slightly differently: "Just a Sprig of Mountain
    Heather." Under the sprigs, which a slit in the cover keeps in place, the booklet
    is identified as a "Souvenir from the Dominion Parks Branch of the Department
    of the Interior Ottawa, Canada." The booklet is bound with fine, tasselled cord.
    The title page bears the names of the department's minister, W.J. Roche, and
    deputy minister, W.W. Cory, as well as that of Harkin. It also bears an italicized
    epigraph—"*Living flowers that skirt the eternal frost.*" This is a portion of the sixty-
    fourth line of Samuel Taylor Coleridge's poem, "Hymn before Sun-rise, in the
    Vale of Chamouni" (1802). Implicitly, this epigraph not only links the Rocky
    Mountain "playgrounds" with one of the best-known tourist destinations in the
    French Alps but also colours them with the imagination of one of the foremost
    Romantic nature poets, thereby cultivating their nature as something wholly
    recognizable and admirable. To ensure that the connection to the Old World is
    not lost on readers, the first of fourteen pages of text clarifies that the very
    heather that is so renowned in Scotland is to be found in a "sunny alpine
    garden" of "Simpson Pass, Rocky Mountains Park," from which the sprig was
    gathered (3).

16 Doyle, *Memories and Adventures*, 306; *Western Wanderings*, 68–69. He was rather more
    candid with a Winnipeg newspaperman, however: "Col. Rogers is working hard
    and doing wonders, but it will take considerable time to get that immense terri-
    tory in shape" (qtd in Roden and Roden, "Introduction," *Western Wanderings*, 28).

17 Benham, "Jasper Park," 4.

18 A further need arose as Bridgland worked. It came from military intelligence as
    realization began to dawn that the Great War was not to be a short war. Jasper
    played a direct role in the war the year after Bridgland's work when it was chosen
    as the location for an internment camp for prisoners of war. Apparently, four-
    teen buildings were erected "on a parcel of low-lying ground between the railway
    station and the Athabasca River." See W.A. Waiser, *Park Prisoners: The Untold Story of
    Western Canada's National Parks, 1915–1946* (Saskatoon: Fifth House, 1995), 33.

19 "The records of the Board of Examiners clarify that those who sought to become
    Dominion Land Surveyors usually had to write a series of examinations. If they
    succeeded in the first set, they would be granted a preliminary certificate. This

certificate allowed them to article with a Dominion Land Surveyor. Depending on their education, their term of articles could be one year, three years, or more. The article papers had to be filed with the Board to be effective in establishing the beginning and duration of the article. Usually, an articling student worked with a number of surveyors in order to gain a wide variety of experience. While thus affiliated, the student would be writing the next two sets of examinations and trying to complete the other requirements for a commission. When he obtained that commission, he became a Dominion Land Surveyor" (A.M. MacLeod, Legislative Advisor to the Surveyor General of Canada Lands, Legal Surveys Division, Natural Resources Canada, Ottawa, to the author, 7 May 2003).

20 M.P. Bridgland, *Photographic Surveying*, Dept. of the Interior Bulletin 56 (Ottawa: F.A. Acland, 1924), 14–15.

21 The entrance to the coal mine and several mine buildings stood near Mountain Creek at the northeast base of Roche Miette, and the housing for the mine workers was perched on a high bench three hundred feet up from the river. From there, the coal miners and their families overlooked the Athabasca and the small sister settlement of Bedson on the other side of the river.

22 Canada, Parliament, "Report of the Commissioner of Dominion Parks, 1915," *Sessional Papers, 1916* (Ottawa: Queen's Printer, 1916), no. 25: 66; Jasper-Yellowhead Historical Society Museum, Jasper: Shovar File; Schools File; Jasper History File; all cited in Brenda Gainer, "The Human History of Jasper National Park," Report Ser. no. 441, typescript report for Parks Canada (1981), 171–72, 205. The population of Jasper today ranges from about 4,750 in the dead of winter to over 30,000 of a mid-summer's weekend.

    Although the permanent population was meagre, Jasper's population was already swelling in the summer, as it does today. As early as 1910, Chief Forest Ranger and Acting Superintendent McLaggan supervised the construction of temporary accommodations, dining facilities, and stables, at eight points in the park. The park's headquarters, now the Information Centre with Heritage designation, were constructed in 1913. In that year, too, a summer cottage subdivision was surveyed near Pyramid Lake. By 1914, forty-six lots had been leased to private enterprises, and stables, trade shops, one school, and two churches had been built. Tent City was up and running in the summer of 1915 (although the war put a stop to it, and its second year of operation did not come until 1919 [Smith, *Jasper Park Lodge*, 9]). So Bridgland arrived at a place where the tourism industry was steadily growing, if not yet booming.

23 Apparently, Horseshoe Lake—a good name for a lake the shape of which resembles that of a horseshoe—was renamed Lac Beauvert in 1915 to increase its appeal to tourists (Great Plains Research Consultants, "Jasper," 250).

24 Smith, *Jasper Park Lodge*, 7.

25 East of the town site, where the main street veers under the railway bridge to connect with Yellowhead Highway, Cottonwood Creek campground was the first to open, in 1927 (Lothian, *Brief History*, 56).

26 "Real Tent City in the heart of the Rockies," *Morning Bulletin* (Edmonton),
     6 May 1916, 17.

27 Bridgland, *Photographic Surveying*, 16.

28 Until 1914, Thumb, not Signal, was the mountain's name. It changed because just
     the year before the arrival of Bridgland's survey crew, early telephone technology
     appeared on the eastern slopes in the form of a single-line ground-return line.
     This involved stringing from poles or trees a line of galvanized wire through
     insulators; thereby, fire ranger stations communicated with one another both in
     the park and in the forest reserves east, north, and south of it.

29 For a discussion of the roles played by the golf course and by Group of Seven
     painters in the development of tourism in Jasper, see I.S. MacLaren, "Cultured
     Wilderness in Jasper National Park," *Journal of Canadian Studies* 34.3 (Fall 1999):
     24–31.

30 Bridgland, *Photographic Surveying*, 16.

31 Readers might well wonder, then, after visiting the website on which all the
     photos of the 1915 season are made available, why only one station's photos from
     Signal Mountain are included, why Signal Mountain is not the first station in
     the succession of ninety-two numbered stations, and why the photograph
     numbers for the eight photographs from Signal are numbered, not from one to
     through eight, but rather from 222 through 229. All reasonable questions, they
     point to the considerable reorganization and selection of the entire season's work
     that Bridgland undertook each year of his career. Sometimes, as well, mountains
     were revisited in the course of the season for a second "shoot" under better
     weather.

     More than three sets of numbers may be found either in lists or along the
     edges of the glass plate negatives themselves. It appears from one set that
     Bridgland reserved numbers 1–299 for his survey team and started Hyatt's team
     at 300, but the correlation is not exact. What readers can assuredly conclude
     however is that the series of station and photograph numbers finalized by
     Bridgland is uninterrupted, and that it is unvarying in the glass plate negatives,
     the references given in captions to photographs later used in his book *Description
     of & Guide to Jasper Park*, and in the numbers used to identify the Jasper season's
     stations and photographs by the Rocky Mountain Repeat Photography Project
     and available on its website at http://bridgland.sunsite.ualberta.ca/jasper/.

32 Édouard-Gaston Deville, *Photographic Surveying, including the Elements of Descriptive
     Geometry and Perspective* (Ottawa: Government Printing Bureau, 1889), 2d ed.
     (Ottawa: Survey Office, 1895).

33 If either of these conditions could not be met, the levels had to be rechecked. If
     they were found to deviate, Bridgland could still level the camera by ascertaining
     where in the graduated tube of the level he had to move the bubble.

34 "Notes by M.P. Bridgland, D.L.S., and A.J. Campbell, D.L.S.," in A.O. Wheeler,
     "The Application of Photography," 91.

35 M.P. Bridgland to É.-G. Deville, 12 Oct. 1915, Surveys Division, Department of the Interior, LAC, RG 88, vol 148, file 15756 gives J.W. as Martin's initials. Surprisingly, this name does not match any in the surviving records of the DLS, which include entries only for Walter Harold Martin, who was articled to C.F. Aylesworth 15 May 1911, and Frederick John Martin, who was articled to W. Christie 18 May 1914.

36 *Description of & Guide to Jasper Park*, 61, 63.

37 The developing supplies used by Bridgland in one year consisted of the following: 6 lbs sulphite of soda, 8 oz pyrogalthic acid, 4 lbs carbonate of soda, 3 oz metol, 3 oz ammonia bromide, 1 oz potassium bromide, 6 oz liquid ammonia, 1 oz citric acid, 12 lbs hyposulphite of soda, and 2 lbs powdered chrome alum (1 lb = 454 g; 1 oz = 28.35 g).

38 Bridgland, *Photographic Surveying*, 15.

39 M.P. Bridgland to É.-G. Deville, 23 Mar. 1916, Surveys Division, Department of the Interior, LAC, RG 88, vol 353, file 15756.

40 M.P. Bridgland, Undated Abstract of Undated Report, LAC, RG 88, vol 148, file 15756.

41 Bridgland, *Description of & Guide to Jasper Park*, 31.

42 At this point in his survey, remote from an already established control point (on the railway lines, for example, or the part of the Athabasca valley that had already undergone township surveys by St Cyr, Hawkins, Herriot, McFarlane, and Matheson), Bridgland likely ran a stadia survey (as described above in Part I, "Measuring the West before Bridgland"). It would have been the fastest way by which to establish some control points for the phototopographical work. Thereby, and at minimum expense of both manpower and money, he could link his survey to already known control points established by earlier surveys.

43 Aphrodite Karamitsanis, *Place Names of Alberta, Vol. 1: Mountains, Mountain Parks and Foothills* (Edmonton: Alberta Community Development, Friends of Geographical Names of Alberta Society, and University of Calgary Press, 1991), 53.

44 Bridgland, *Description of & Guide to Jasper Park*, 51.

45 "Diary of M.P. Bridgland, D.L. Surveyor," 21 June to 18 Oct. 1915, Geomatics Canada, Earth Sciences Sector, Legal Surveys Division, 14929; copy courtesy Alec M. MacLeod. Brief throughout, the printed form diary contains only fifteen other words for the entry under 29 July. They precede the three-word report of Hyatt's death: "Took day to straighten up things in camp + get ready for the next trip." This may be regarded as the official diary of the field season; of course, there were field diaries, much messier, notational, and occasionally misspelled or ungrammatical earlier versions of this fair copy, one kept by Bridgland, the other by Hyatt and after his death by his replacement, Norfolk. These are housed in Special Collections, Cameron Library, University of Alberta. Under the wrong name, an Edmonton newspaper reported the "sad drowning accident...when Edward Wyatt of the Topographical department, Ottawa, was

out swimming and was suddenly overcome by cramps and unable to reach the shore sank in deep water" ("Drowns near Jasper," *Edmonton Journal* 31 July 1915, 5).

46 In advance of the transfer of resources to the provinces in 1930 and the inauguration of a new national parks act in which resource harvesting activities such as mining and lumbering would for the first time be prohibited, the 329 mi^2 (852 km^2) cut off on either side of the Athabasca River was proposed in 1929, at the same time as several other parcels of land were lopped off the eastern borders of the mountain parks. The largest of these was the 630 mi^2 (1,631 km^2) of Banff Park that included the Kananaskis Range and the Front Range (in the area of the mining activity at Canmore and Exshaw) as far north as Ghost River. See *Map showing Areas which it is Proposed to withdraw from Rocky Mountain and Jasper Parks*, 1:792,000 (Ottawa: Dept. of the Interior, 1929).

47 M.P. Bridgland to É.-G. Deville, Surveys Division, Department of the Interior 11 Aug. 1915, LAC, RG 88, vol. 142, file 13913.

48 Bridgland, *Description of & Guide to Jasper Park*, 31.

49 This marking of the name is singular because Bridgland's records were otherwise very orderly. The upper corners of his pages, for example, were routinely filled with numbers from his readings, notes on exposure times, and light conditions.

50 Jeffrey S. Murray, "Mapping the Mountains," *Legion Magazine* 76:3 (May/June 2001), 47; also available electronically at http://www.legionmagazine.com/features/canadianreflections/01-05.asp

51 A.O. Wheeler, M.P. Bridgland, and A.J. Campbell, "The Application of Photography," 86.

52 Bridgland, *Photographic Surveying*, 33–34.

53 É.-G. Deville to M.P. Bridgland, 5 Dec. 1916, Surveys Division, Dept. of the Interior. LAC, RG 88, vol 148, file 15756.

54 For more on this idea, see Gillian Rose, *Visual Methodologies: An Introduction to the Interpretation of Visual Materials* (Thousand Oaks, CA: Sage Publications, 2001).

55 Paul Berger, "Doubling: This Then That," in *Second View: The Rephotographic Survey Project*, ed. Mark Klett, Ellen Manchester, and JoAnn Verburg (Albuquerque: University of New Mexico Press, 1984), 45–52 (45 qtd); Jenaya Webb, "Imaging and Imagining: Mapping, Repeat Photography and Ecological Restoration in Jasper National Park," MA Thesis, University of Alberta, 2003, 54.

56 Bridgland, *Photographic Surveying*, 15.

57 Dennis Reid, "Photographs by Tom Thomson," National Gallery of Canada, *Bulletin*, no. 16 (1970): 2–36; text available electronically at http://collections.ic.gc.ca/bulletin/num16/reid1.html

58 M.P. Bridgland, "Abstract of Report." n.d., Surveys Division, Dept. of the Interior, LAC, RG 88, vol 148, file 15756. Wheeler claimed in 1920 that in "three months of fair weather, from 500 to 1,000 square miles of country can be mapped by one party, according to the degree of accuracy and detail desired" (A.O. Wheeler, M.P. Bridgland, and A.J. Campbell, "The Application of Photography," 86).

59 É.-G. Deville to Director of Forestry, 8 Mar. 1917, Dept. of the Interior, LAC, RG 88, vol 361, file 16533.

60 "Notes by M.P. Bridgland, D.L.S., and A.J. Campbell, D.L.S.," in A.O. Wheeler, "The Application of Photography," 90.

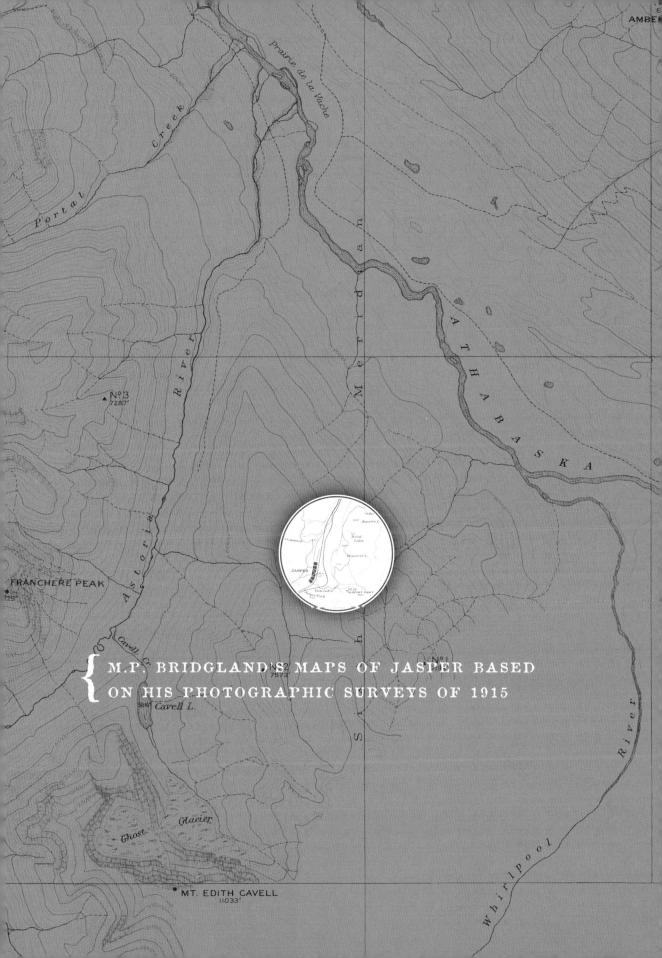

{ M.P. BRIDGLAND'S MAPS OF JASPER BASED
ON HIS PHOTOGRAPHIC SURVEYS OF 1915

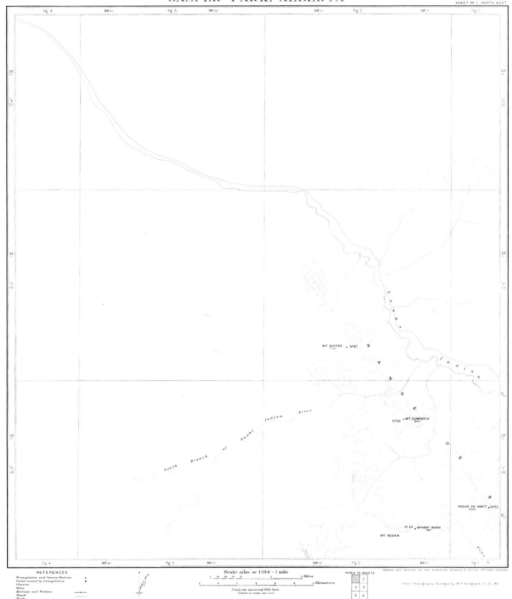

Map of the Central Part of Jasper Park, Alberta (1916), Sheet One, Northwest

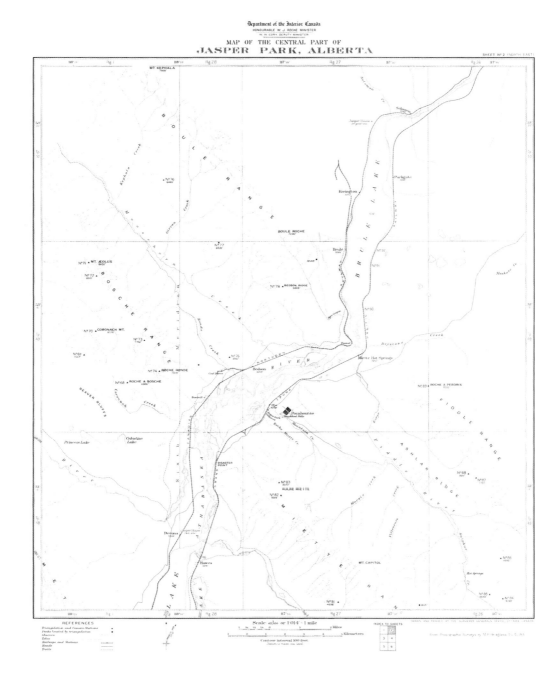

Map of the Central Part of Jasper Park, Alberta (1916), Sheet Two, Northeast

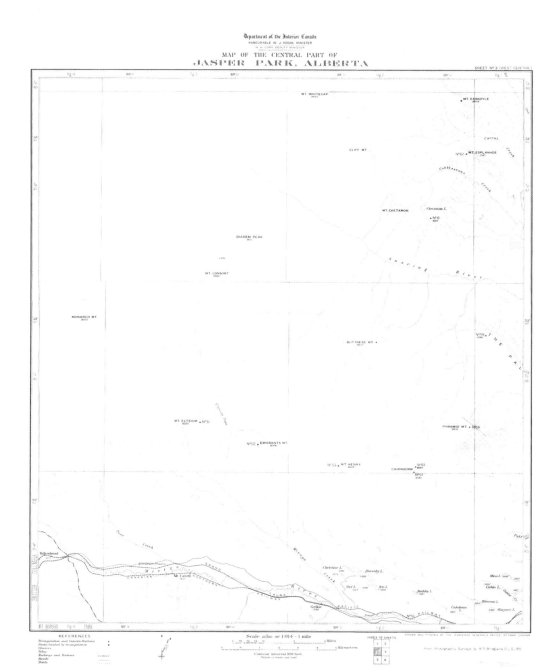

Map of the Central Part of Jasper Park, Alberta (1916), Sheet Three, West Central

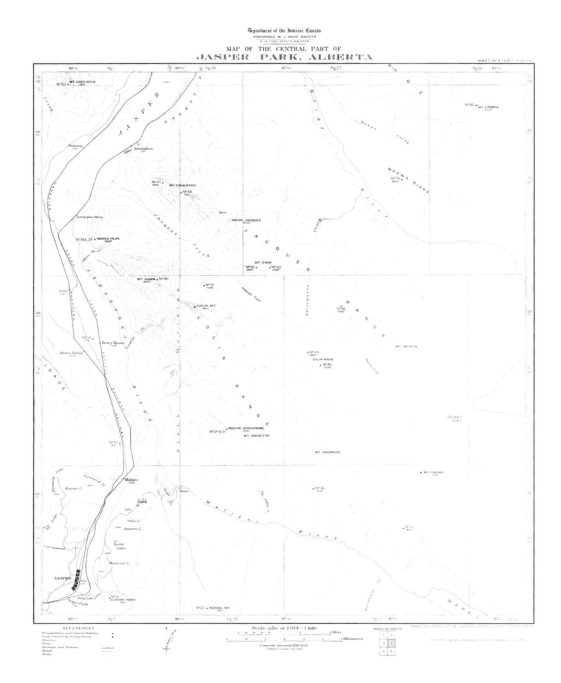

Map of the Central Part of Jasper Park, Alberta (1916), Sheet Four, East Central

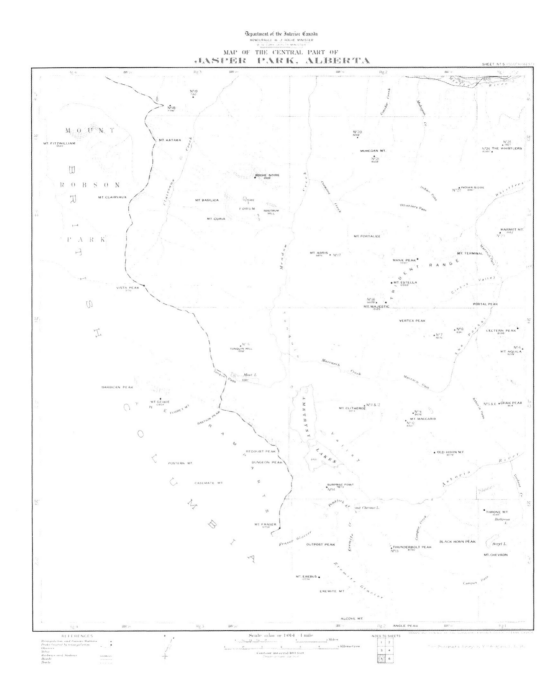

Map of the Central Part of Jasper Park, Alberta (1916), Sheet Five, Southwest

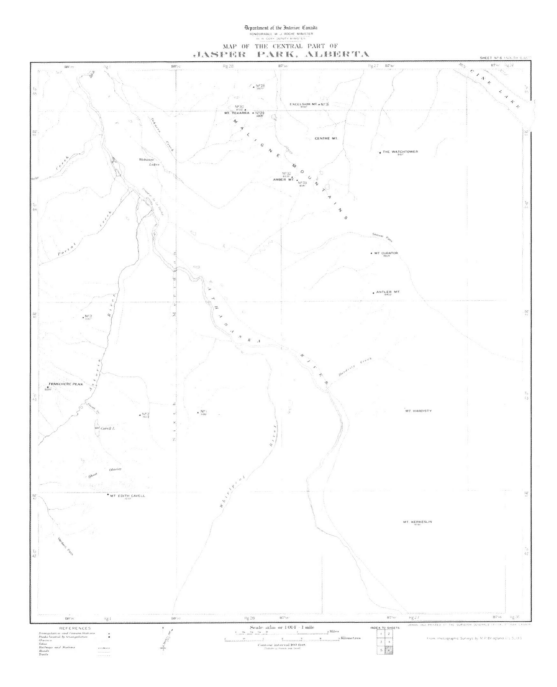

Map of the Central Part of Jasper Park, Alberta (1916), Sheet Six, Southeast

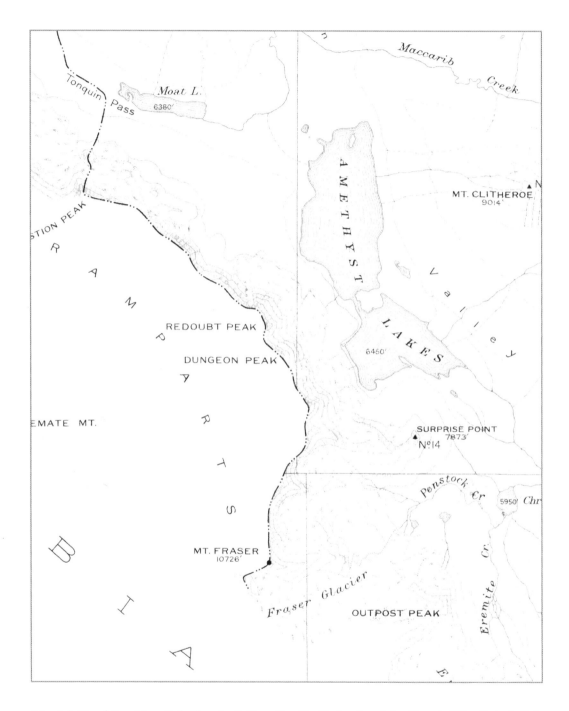

Map of the Central Part of Jasper Park, Alberta (1916), Sheet Five (detail): Amethyst Lakes, Tonquin Valley, Redoubt Peak and Dungeon Peak (The Ramparts), Surprise Point

etailed National Topographical Survey maps with which
we are familiar today make the six sheets of M.P. Bridgland's
ninety-year-old *Map of the Central Part of Jasper Park, Alberta* look
scanty in their detail, but one's admiration for this map increases with
the realization that his was the first systematically made map of any
portion of what we know today as Jasper National Park. To be sure,
earlier railway and township surveying had resulted in maps of the
floor of the Athabasca River valley, but the region generally was cartographically unrecorded, and certainly neither railway nor township
surveyors had scaled the park's mountains as part of their map-making
efforts. The detail—one-hundred-foot contour intervals and a scale of
just over a mile to the inch—was unprecedented.

To achieve such maps, Bridgland and his crews had to come to
know the central part of Jasper Park intimately in fewer than one
hundred days. The 735 photographs that he, Hyatt, and Norfolk shot
made that familiarity possible. So when he sat down at the DLS office
in Calgary every morning of the winter of 1915–16, he could proceed to
map-making with confidence. But surely with some sadness, too, for
his collaboration had come to an end with the very capable young Ed
Hyatt, whose phototopographical surveys had also helped make
possible the first map of the Crowsnest Forest Reserve and Waterton
Lakes Park earlier in 1915.

With Hyatt dead in Jasper and buried in Revelstoke, Bridgland
pressed on. His next challenge was an entire book, *Description of & Guide
to Jasper Park,* which both the Dominion Lands Survey and the
Dominion Parks Branch wanted published as quickly as possible. The
six sheets of the *Map of the Central Part of Jasper Park* could be purchased
with that guide just as soon as it appeared in 1917. In mid-career, the
Mapper of Mountains had hit his stride. ✳

Postern Mt.

Mt. Geikie
Turret Mt.
Bastion Pk.   Barbican Pk.

Moat Lake and
Tonquin Pass

Tonquin Hill

{ HISTORY OF THE REGION AND OF ATTEMPTS AT CLIMBING PARTICULAR
MOUNTAINS COMBINE WITH EXPRESSIONS OF AESTHETIC APPRECIATION
FOR THEIR FORMS, AND INFORMATION FOR THE CASUAL TOURIST.

FOUR

Bridgland's Life and Times
1916–1948

✳

M.P. Becomes an Author, 1916

The Jasper field season was a remarkable one. Its scope was unparalleled but so too was its tragic loss of a promising young surveyor. In terms of the daily work, though, it was just another summer's work for the dedicated Mapper of Mountains. Bridgland produced another outstanding map from the Jasper survey, the one destined for tourists that derived from his DLS map. This creation occupied him in the summer of 1916, a rare one for him because he stayed in the office. Yet, it was that summer that also enabled him to make a beginning on the ninety-seven-page illustrated booklet for tourists, entitled *Description of & Guide to Jasper Park,* which he co-wrote with Robert Douglas (b.1881), secretary of the Geographic Board of Canada.[1]

THE PUBLIC DEMAND
FOR THE BOOK
WAS STRONG AND
ENDURING.

*T*he Department of the Interior first published the guide in 1917, with a print-run of 3,000 copies. In its opening pages, along with the tourist map that faced the title page, was advertised Bridgland's DLS map in six sheets. It could be ordered for fifteen cents per sheet, or mounted and dissected for the pocket at fifty cents per sheet. No similar public sale occurred with his other DLS maps. The public demand for the book was strong and enduring. Very few detailed DLS maps were as carefully made as Bridgland's or offered so widely for public sale with an accompanying publication.[2] Needless to say, Surveyor General Deville and Parks Commissioner Harkin teamed up to ensure that all avenues for publicity be exploited. Not surprisingly, the book opens with Sir Arthur Conan Doyle's poem, "The Athabaska Trail":

My life is gliding downwards; It speeds swifter to the day
When it shoots the last dark canon to the Plains of Far-away,
But while its stream is running through the years that are to be,
The mighty voice of Canada will ever call to me.
I shall hear the roar of rivers where the rapids foam and tear,
I shall smell the virgin upland with its balsam-laden air,
And shall dream that I am riding down the winding woody vale,
With the packer and the packhorse on the Athabaska Trail.

I have passed the warden cities at the Eastern water-gate,
Where the hero and the martyr laid the corner-stone of State,
The habitant, coureur-des-bois—and hardy voyageur.
Where lives a breed more strong at need to venture or endure?
I have seen the gorge of Erie where the roaring waters run,
I have crossed the Inland Ocean, lying golden in the sun,
But the last and best and sweetest is the ride by hill and dale,
With the packer and the packhorse on the Athabaska Trail.

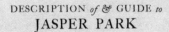

DESCRIPTION *of & *GUIDE *to*
JASPER PARK

OTTAWA
DEPARTMENT OF THE INTERIOR
1917

Title Page, *Description of & Guide to Jasper Park* (1917).

I'll dream again of fields of grain that stretch from sky to sky,
And the little prairie hamlets where the cars go roaring by,
Wooden hamlets as I saw them—noble cities still to be
To girdle stately Canada with gems from sea to sea;
Mother of a mighty manhood, Land of glamour and of hope,
From the eastward sea-swept Island to the sunny Western slope,
Ever more my heart is with you, ever more till life shall fail,
I'll be out with pack and packer on the Athabaska Trail.

ARTHUR CONAN DOYLE

Jasper park, Alberta, June 18, 1914[3]

That Doyle had conceived of the entire Dominion's majesty as being centred on Jasper offered the publicists everything they could have wished as a promotional introduction to the park. The pride of Canada, certainly at its acme during the war years, is at its most fervent at the heart's homeland, the Athabasca Trail. The text takes its cue from Doyle and gives its reader a sense of a deep, resonating heritage in Jasper. "Jasper Park is historic ground," the book begins: "More stirring scenes in the upbuilding of Canada have been staged in it than in any other part of the Rockies."

In an effort to make the area as accessible and attractive as possible to tourists, the authors gave detailed instructions for long and short trips, mostly based on the trails and passes Bridgland had used in his survey. They included not only the information required to find specific trail heads or mountain passes, but also historical and cultural information to help visitors orient themselves to the heritage of the park. In almost every section of the book, history of the region and of attempts at climbing particular mountains combine with expressions of aesthetic appreciation for their forms, and information for the casual tourist. In this way, the book achieves two purposes simultaneously: it echoes the romantic language used in the era's railway pamphlets, which were beckoning the tourist to come to the wilderness while it was still "unknown"; as well, though, it used its role as an official government publication to disseminate authoritative information in word and image about the park. Both publicity and functional information thus combine to reveal the park's majestic attributes.[4]

Just a portion of the extensive treatment accorded Mt Edith Cavell exemplifies well what it was that made the book so interesting. You can hear the history buff, the fellow member of the Alpine Club of Canada, the visitor who just wants a hike, and the tourist who will wait until a wagon-bearing road from Jasper is completed all being addressed in succession:

> To the south, the dominating feature of the landscape is mount Edith Cavell, its massive top frequently clothed in clouds. … Almost due south of Jasper [town site] and about fourteen miles distant in a straight line, [this] massive snow-crowned mountain rises high above the surrounding peaks, its white summit sometimes glistening in the sunlight, but often lost among the clouds. Below, some hanging glaciers showing white

against its dark precipitous rocks, discharge their masses of snow and ice into a large glacier just visible in the valley beneath. This is mount Edith Cavell, named in honour of the heroic nurse who was murdered by the Germans in the fall of 1915.

In the cirque between the mountain and the shoulder extending to the north, there is a hanging glacier about one-third of a square mile in area. This glacier drains into a large one about five hundred feet below by a narrow fan-shaped icefall not more than three hundred feet wide at the top. The lower glacier is of irregular shape extending along the foot of the cliffs for more than a mile, and its greatest width is a little over half a mile. The whole forms the glacier of the Ghost. A well-defined medial moraine marks the junction of the two parts of the ice field. A short distance below the snout of the glacier, nestling in the green timber, lies a small lake about one thousand yards long and three hundred yards wide, which has been named Cavell lake, while the stream flowing from it to Astoria river has been called Cavell creek.

It is doubtful if the mountain can be climbed from this side, but from a camp at the lake, the glaciers and the valley above could be examined. The same camp would make a convenient base from which to explore the long ridge extending eastward toward the Athabaska valley. From points on this ridge, only 1,500 to 2,000 feet above the lake, splendid views may be obtained of Athabaska river looking either up or down, of the lower part of the Whirlpool or of the valley of Astoria river. Opposite the mouth of Astoria river on the east side of the Athabaska is seen the prairie de la Vache, of which [fur trader] Ross Cox [1793–1853] says [in *Adventures on the Columbia River* (1831)]:

> From the junction of the two rivers to the old fort, the country on each side presents a pleasing variety of prairies, open woods and gently rising eminences; and one spot in particular, called la prairie de la Vache (in consequence of buffalo having formerly been killed in it) forms a landscape that for rural beauty cannot be excelled in any country.

… As might be expected a mountain so prominent as Mt. Edith Cavell could not fail to attract the attention of moun-

IN THE CIRQUE
BETWEEN THE
MOUNTAIN AND THE
SHOULDER EXTENDING
TO THE NORTH, THERE
IS A HANGING GLACIER
ABOUT ONE-THIRD OF A
SQUARE MILE IN AREA.

taineers. Thus Principal [of Queen's University] Grant writes [in *Ocean to Ocean* (1873)]:——

> At the end of lake Jasper, a strath from two to five miles wide, which may still be called the Jasper valley, bends to the south. Our first look up this valley showed new lines of mountains on both sides, closed at the head by a great mountain so white with snow that it looked like a sheet suspended from the heavens. That, Valad [the Métis guide] said, was "La montagne de la grande traverse," adding that the road to the Columbia country up the formidable Athabaska pass, lay along its south-eastern base, while our road would turn west up the valley of the river Myette. He mentioned the old local titles of the mountains on this side, but every passer-by thinks that he has a right to give his own and his friends' names to them over again.

THE MOUNTAIN OF THE GREAT CROSSING WAS ONE OF THE GUIDEPOSTS OF THE EARLY TRAVELLERS; IT MARKED THE PLACE WHERE THE ATHABASKA HAD TO BE FORDED AND THE CONTINENTAL DIVIDE CROSSED.

The mountain of the Great Crossing was one of the guideposts of the early travellers; it marked the place where the Athabaska had to be forded and the continental divide crossed. Sir James Hector [1834–1907, a member of the Palliser expedition] named it mount Le Duc [in 1858], and in 1911, A.O. Wheeler named it mount Fitzhugh. Locally, it was known as mount Geikie, which was a mistake, the real mount Geikie being some twelve miles farther west. In 1913, shortly after the construction of the railway, an attempt to reach the summit was made by A.L. Mumm and G.E. Howard, both members of the English Alpine Club, but owing to unfavourable weather the attempt failed, although Mr. Mumm, accompanied by his guide, ascended to within four hundred feet of the top. In 1915 another and a successful attempt was made by Professor E.W.D. Holway and Dr. A.J. Gilmore. The ascent was made by the west arête from a camp on the wide grassy pass south of the mountain, between Astoria and Whirlpool river. Professor Holway says no great difficulty was encountered, though there was much step-cutting part of the way up very steep slopes.

A good trail has been built from Jasper to Cavell lake, a distance of about fifteen miles. This trail follows the wagon road across the Miette river and then turns south along the base of the Whistlers. Passing along the flats on the west side of Athabaska river, it crosses Whistlers creek, Portal creek and Astoria river, and then turns up the last, gradually ascending along the side of the low ridge on the south until the lake is reached. Soon after turning up Astoria river, the trail passes close to a gorge on the main stream. Most of the way the trail is along open flats or through old brulé but for the last two miles the mountain side is covered with green timber. In addition to the trail, a good road is being constructed to this point.[5]

Even a touch of Bridgland's dry humour seems to find a place in this comprehensive passage: After quoting how Grant's guide scorned every passer-by's penchant for bestowing names on mountains, he proceeds to tell us that, among others, Wheeler, his difficult mentor, committed the same self-indulgence in 1911! But the quotation of Cox reminds us of the fur trade era one hundred years before, and of Grant the early railway surveys fifty years before. The history is brought right up to date by both the notice of alpinists' efforts to climb Edith Cavell and indeed of the new name for the mountain, which even as recently as 1915, Bridgland had not used in his survey.

The book's appendices make plain its intended audience. Trips from Jasper town site or Pocahontas ("Poco") are measured out with Bridglandian care, and include distances along the way to particular intermediate points on any route and the number of days required for each trip. (All M.P.'s experience with the ACC summer camps comes to bear on his knowledgeable contribution of such information.) Mountains are listed alphabetically, followed by their elevation and the source of their names is provided. As well, an index gets the reader quickly to the point in the book where each mountain or other feature receives description and discussion. Interestingly, as much attention is paid to trips available from Pocahontas as from Jasper. In a guide for visitors in the decades before the arrival of the automobile tourist, the idea of the park's focal point being Jasper town site itself does not come across. And Pocohantas was a thriving little community still in 1917, poised to add to its mining operation the role of coordination point between its Grand Trunk Pacific railway station and the expected road leading from the Athabasca valley up to Miette Hot Springs and its grand hotel, which would remain a phantom.

IN A GUIDE FOR VISITORS IN THE DECADES BEFORE THE ARRIVAL OF THE AUTOMOBILE TOURIST, THE IDEA OF THE PARK'S FOCAL POINT BEING JASPER TOWN SITE ITSELF DOES NOT COME ACROSS.

Description of & Guide to Jasper Park, Commissioner Harkin noted in 1920, had "been of great value and [was] in constant demand,"[6] retailing for fifty cents, except on the trains, where it sold for an extra dime. As mentioned earlier, the booklet differs markedly from A.O. Wheeler's compendious two-volume *The Selkirk Range* (1905). That work is not a particular park guide and is not particularly directed towards the tourist. Only after the histories of exploration and surveys of the Selkirks does Wheeler move to a section that might be expected in a park guide, and it is directed to alpinists. In *The Selkirk Mountains: A Guide for Mountain Climbers and Pilgrims*, published in 1912 (after he had resigned from the DLS) and not published by the government's printing office, Wheeler and his co-author, ACC past-secretary Elizabeth Parker, offered something more akin to what Bridgland and Robert Douglas would develop five years later, but still, as its subtitle clarifies and indeed as the region covered would suggest, there is little in the book to attract visitors other than those looking for mountaineering adventures.[7] Altogether, *Description & Guide* is a more catholic endeavour, directed at a far wider readership. And it is a celebration of the phototopographical method, which, by the end of the century's second decade, would survey and map about 40,000 mi² (104,000 km²) of the Dominion.[8]

A guide by a phototopographer was bound to be amply illustrated. Most of the landscape photos were selected from M.P.'s topographical work. But some he must have taken with such an illustrated guide evidently in mind. Deville took it upon himself to integrate photos and text, and to edit the whole. This was perhaps a poor choice, for the photos could have been better coordinated with the text than they often are. Indeed, readers are left scratching their heads at some points. Perhaps Deville had never visited Jasper or, if he had, never set foot outside of town, and was arranging the work without the intimate knowledge that Bridgland possessed. This difficulty with the book is countered by the provision of Bridgland's maps, which could always give readers a steadying influence when the text or photos caused them to lose or question their bearings.

One photo of Mt Edith Cavell has lost none of its allure. It was taken from Mt Maccarib, a viewpoint seen by few tourists even today. As well, there is a superb photograph shot while looking south from Edith Cavell and showing the upper valley of the Athabasca River in the background and the lower Whirlpool River—entry to Athabasca Pass and the Columbia River watershed—in the foreground. This illustrates exactly

AND IT IS A
CELEBRATION OF THE
PHOTOTOPOGRAPHICAL
METHOD, WHICH, BY
THE END OF THE
CENTURY'S SECOND
DECADE, WOULD
SURVEY AND MAP
ABOUT 40,000 MI²
(104,000 KM²) OF
THE DOMINION.

Photograph by M.P. Bridgland, La Montagne de la Grande Traverse (Edith Cavell) and Verdant Valley from Mt Maccarib, Stn 9, no 76, direction eastsoutheast, 1915. From M.P. Bridgland *Description of & Guide to Jasper Park*, 43. Courtesy Jasper National Park.

Photograph by M.P. Bridgland, Athabasca and Whirlpool rivers from Cavell Meadows, Stn 1, no 12, direction southeast, 1915. From M.P. Bridgland *Description of & Guide to Jasper Park*, p. 53. Courtesy Library and Archives Canada R214.

Postern Mt.

Mt. Geikie
Turret Mt.
Bastion Pk.   Barbican Pk.

Moat Lake and
Tonquin Pass

Tonquin Hill

The Ramparts and Tonquin Pass from Mt. Clitheroe (11)
Amethyst lakes in foreground. The continental divide crosses the pass a few hundred feet beyond Moat lake

Photograph by M.P. Bridgland, The Ramparts, Amethyst Lakes, and Tonquin Valley from Mt Clitheroe, Stn 11, no. 89, direction west, 1915. From M.P. Bridgland *Description of & Guide to Jasper Park*, 48.

Photograph by M.P. Bridgland, The Ramparts, Amethyst Lakes, and Tonquin Valley from Mt Clitheroe, Stn 11, no. 89, direction west, 1915. Courtesy Library and Archives Canada R214.

the word picture that forms part of the text about Mt Edith Cavell, and it leaves one wishing that the entire book were so well linked. Many of the photos showing an array of peaks have each of them identified above the picture. Illustrations of flowers and animals grace the volume, as well, and some shots identify particular buildings, such as the famous park superintendent's residence (completed in 1914), located across Connaught Drive from the VIA/CNR train station.

One detail about Jasper in its early years that *Description & Guide* does not shy away from mentioning, despite the book's purpose, is the frequency with which fire had burned the region: "Much of the area within the park has been swept by fire, though fortunately the heads of many fine valleys have been spared." And the authors speak of the "denuded nature of the country," as well as the frequency with which brûlé has to be negotiated as one follows trails to particular sights or peaks.[9] That this subject is mentioned at all suggests the confidence with which a writer in 1917 anticipates a change in the landscape now that it had been brought under the designation of a park. Indeed, the text uncharacteristically goes out of its way to blame fires on a "quarrel between two tribes of Indians," a quarrel that, with the removal of residents from the Athabasca valley in 1910, one infers could not be repeated. But *Description & Guide* is very much a book of its times in its relegation of Native people to the past. The idea that they were vanishing had long held sway in North America, feeding the Myth of Progress a steady nutritional story that the myth needed to justify itself. Of course, the expulsion of the Métis homesteaders and the prohibition against hunting made the disappearance a fact in Jasper Park.

Inspecting the photos or reading the accompanying text immediately gives you the feeling of being in the presence of the best possible guide to the area, even though it was hardly familiar ground for Bridgland. But the authoritative rehearsal of history instills confidence. It is no wonder that copies of *Description of & Guide to Jasper Park* are very difficult to find today, or that they cost well over $100 when they do surface for sale. Exceptionally popular, the guide went through several printings.

Photograph by M.P. Bridgland of a bear in tree. The book's caption reads, "Dinner Ready? Wild animals in the park become very tame. This bear was quite punctual at the meal hours of the survey party." From M.P. Bridgland *Description of & Guide to Jasper Park*, 34.

Photograph by M.P. Bridgland of Henry House Flats, Athabasca River Valley, Colin Range, Stn 58, no 467, direction eastnortheast, 1915. Compare the scant presence of conifers between the flats and the river with the quantity present today. Courtesy Library and Archives Canada R214.

Photograph by J.M. Rhemtulla and E.S. Higgs of Henry House Flats, Athabasca River Valley, Colin Range, Stn 58, no 467, direction eastnortheast, 1998. Courtesy Rocky Mountain Repeat Photography Project, University of Alberta, University of Victoria.

Moreover, it appears that this publication prompted a campaign of similar guides for the other parks: Mabel Bertha Williams, secretary to Harkin who remained parks commissioner for a full quarter-century (1911–1936), went on in the ensuing three decades to publish a dozen such guides for both mountain and other national parks. Even today, judging by what is handed to every visitor upon entry into any of Canada's Rocky Mountain national parks, the work of Bridgland, Douglas, and Deville serves as the template for the mountain guides.[10]

✳ *Bow River and Clearwater Forest Reserves, and Beyond,* All the Way to Pitt Lake, 1917–1921 ✳

DESPITE HIS SKILL AS A GUIDE IN PROSE AND PICTURE, Bridgland was probably elated to return to the mountains in the summer of 1917 after twenty months in the office. His work for the next four years would lie in the Bow River and Clearwater forest reserves, along the valleys of the eastern border of Banff Park. In the second year after resuming his summer surveys, he hired as his assistant Calgarian Ley Edwards Harris (1890–1983). Like M.P., he had grown up in Ontario (Gananoque) and come West to work. Harris continued in association with him for the next thirteen years. In two interviews conducted in the 1970s by Lizzie Rummel, a rancher and famous mountaineer in her own right, Harris confessed that there were innumerable "experiences and incidents" during his first few years in the mountains.[11] Although several of these incidents may have given Bridgland reason to doubt Harris' abilities as a mountaineer, he never made any disparaging or discouraging comments. Encouraged, Harris persevered.

LEY ARRIVED AT HIS FIRST CAMP WITH NO CLIMBING EXPERIENCE AND PROBABLY NOT THE FOGGIEST IDEA OF HOW TO SCALE A ROCK FACE WITH THIRTY POUNDS OF EQUIPMENT IN A WOODEN BOX STRAPPED TO HIS BACK.

Harris passed his preliminary DLS examinations in 1910 on the second attempt, and, after articling with other DLS surveyors, wrote and passed his qualifying examinations in 1918, receiving his DLS commission on 1 March of that year, shortly before Bridgland took him into the mountains for the first time.[12] As had M.P. sixteen years earlier, Ley arrived at his first camp with no climbing experience and probably not the foggiest idea of how to scale a rock face with thirty pounds of equipment in a wooden box strapped to his back. For his first climb, his mentor treated him rather differently than Wheeler had treated Bridgland. After all the experience M.P. had gained in helping Alpine Club members achieve their first climbs, he was considerate, at least as Harris remembers those first

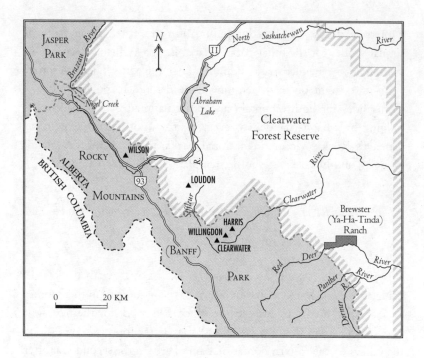

Bridgland's Surveys (with Harris), 1917-1920

THAT SUMMER'S WORK
WOULD THUS REMAIN
INDELIBLY ETCHED IN
HARRIS' MEMORY.

days, leading the younger man over the difficult passages, advising him how to place his feet and hands to gain the best grip, and how to keep his centre of gravity close to the rock as he climbed.

At the station that they built on the summit, Harris remembers how picturesque were the shadows cast by the clouds on the Red Deer River. The wonderful sight of the mountains and the valleys filled him with awe for the landscape, but this feeling was soon overshadowed by his realization that he was to be turned loose for the rest of the summer. The very next day, he found himself climbing on his own and leading his own assistant. Chief of Party Bridgland always set the example for efficient and safe climbing but such a short work season precluded teaching the subtle nuances of technique. That summer's work would thus remain indelibly etched in Harris' memory: assuming such responsibility in such landscape was an awe-inspiring experience in every sense. Every crease in the upper Red Deer and Panther river valleys as well as what is now called Dormer River became his country, it seemed.[13]

Throughout the summer, as he did every season, Bridgland had to develop plates on occasion to make sure his photographs were as accurate as he could make them. Unfortunately, the plates he used in June and July were old and did not develop as well as expected. When Deville examined them he sent a new development outfit to him for his future darkroom work. The assemblage of new developing powders, salts, acid fixers, thermometer, and hard rubber trays allowed for far more refined developing than could be achieved with the bottled developing emulsions Bridgland had used in his early years. However, because of the large amount of space necessary for the trays, the silk darkroom tent had to be suspended in a relatively tall tent or cabin, or from a tree branch higher than could be reached from the ground. Before entering the light-proof canopy, he prepared his materials and memorized exactly where he placed them. He opened carefully the small envelope of developing powder and smoothed the contents with the end of a lead pencil, ensuring there were no lumps. He then poured it through a paper cone into a bottle containing a small quantity of lukewarm water. Once fully dissolved, the developer could be further diluted with cold water. With the aid of a small thermometer, M.P. checked that the temperature of the final solution was 65°F (18°C). He placed the bottle of developer in front of an empty tray and then filled a second rubber tray with clean water. Filling the final tray with 115 grams of acid fixer, he then closed the tent around himself and his materials, tucking in the edges and seams to prevent light from leaking in.

FILLING THE FINAL TRAY WITH 115 GRAMS OF ACID FIXER, HE THEN CLOSED THE TENT AROUND HIMSELF AND HIS MATERIALS, TUCKING IN THE EDGES AND SEAMS TO PREVENT LIGHT FROM LEAKING IN.

The next step was to take the glass plate negative out of its holder, place it in the empty tray, pour the developer over it, and gently rock the tray for six full minutes, ensuring that the developer remained in motion. After this time, he rinsed the negative in clean water, then placed it in the fixer for three to five minutes. The results, although painstaking to produce, were highly satisfactory. Except for the few Bridgland developed in the field, he sent the exposed negatives to Ottawa for developing, but now he could send them confident and satisfied that they met his standards. Great expense was incurred in the shipping of the negatives, but the Department of the Interior insisted on the consistency that the printing of all survey prints only by the same technicians in the same facility could ensure. Once when he was assigned to ship a season's worth of plates to Ottawa by train, Harris decided to ensure the negatives for $3,000, more than his annual salary, because if they disappeared or perished the department would indeed have to pay to have them all retaken.

In 1919, Harris did his best to prove to the chief that he had learned from his first year's experience. When M.P. mentioned that he needed a packer to fetch the horses from Morley, Alberta, Harris volunteered for the job. "I think I can throw a diamond hitch," he said confidently, although anyone who has ever tried to tie this complicated harness knows it's not a skill mastered without a great deal of practice. Bridgland was not the kind of man to argue with a volunteer eager for work, however; he replied, "Very well, you can get the horses from Morley and I'll meet you at the ranger station on the Red Deer." Harris' first attempts to balance the heavy saddlebags, arrange the ropes, and tie the diamond hitch must have met with indifference from the horses, but securing the packs was only the start of his ordeal. The horses were small and usually good-natured about having the 160–200-lb (70–90-kg) loads strapped to them. One horse in particular, however, a bay named Tom, was only occasionally used for packing, and was often unpredictable when he was made to carry any load other than a solitary rider. As Harris led the party cross-country to meet Bridgland, Tom grew increasingly restless. On the second morning of the trip, he suddenly took off at a full gallop, banging his side packs against the trees and bucking wildly. This drama tested Harris' diamond hitch; it held. Unluckily, however, this marked one time when failure would have served him better: the trail took a sharp turn around a deep pool, but Tom left the path and veered into it. He could not extricate himself, and Harris had to cut off the packs, tie a rope around him, and secure the other end around the belly of one of the mares. At first, he tried gently rocking Tom forward and backward to encourage him to climb out of the hole, but when this approach failed he resorted to pulling and yanking. Tom finally emerged, sodden and shaking like a leaf. Harris chased him around to dry him off. When they had both regained their composure, he reset the packs and they resumed the journey. He admits that he never bothered mentioning the incident to Bridgland when he finally reached the ranger station. M.P. probably could have guessed what had happened.

Later that summer, Bridgland and Harris climbed together up a peak that neither of them would soon forget. Mt Harris (10,842 ft [3,299 m]), as Bridgland named it, sits three km northeast of Mt Willingdon on the south side of Clearwater Pass. M.P. found a good place to camp in the timber on the uppermost slopes of the pass. From it, he and his men would survey the surrounding area before proceeding farther down the Siffleur River valley. The first attempt failed. Hiking along the edge of the timberline, he came to a rocky promontory where he had to climb around the edge of an intervening ridge into the

HIKING ALONG THE EDGE OF THE TIMBER-LINE, HE CAME TO A ROCKY PROMONTORY WHERE HE HAD TO CLIMB AROUND THE EDGE OF AN INTER-VENING RIDGE INTO THE HANGING VALLEY BELOW, AND THEN FOLLOW ALONG A ROCK WALL SOUTH OF THE PEAK TO NEAR THE EAST RIDGE.

hanging valley below, and then follow along a rock wall south of the peak to near the east ridge. After all this rather technical climbing, heavy clouds rolled up and discouraged the crew from attempting to summit the peak. Before turning back, however, they ascended a chimney and found that there was no further difficulty beyond it.

A few days later, the group renewed its attempt for the first ascent, this time accompanied by Harris. Although they had moved their camp a few miles downstream in order to eliminate some of the approach time, they still had some very technical rock faces to negotiate, and the weather was not much clearer than on the previous attempt. As they planned on occupying at least two camera stations at the summit, Harris suggested that they take two transits in order to accomplish the work more quickly, but the extra instrument hindered more than helped. They approached a twenty- to thirty-ft rock face, and Bridgland climbed up first, taking a long rope with him. He then dropped one end of the rope down for the crew to tie up the equipment, and hoisted up to safety all seventy pounds of camera, plates, and transits. The very smooth rock offered few handholds. Both Harris and Jack Brinkman, a labourer who accompanied them that day, were hesitant about having to climb it. Reluctant to delay their progress any longer, M.P. agreed to wait at the top of the rock face, anchoring the rope for the two other men to use.

With the weather threatening, there was still part of a glacier to negotiate before they could reach a rocky ridge, and then a narrow chimney that would lead them to a spot sufficiently level for camera stations. Only M.P. had bothered to put some small nails in the soles of his boots, so he alone was prepared for the ice work. He took the necessary photos at the first station, then left Harris to read the angles while he and Brinkman occupied a second station on the north side of the peak where they could see Cataract Creek. Harris busied himself by building a cairn at the site of the first camera station, but, before he noticed, a snowstorm arrived to turn everything around him white. Interminable minutes later, he heard on the shale the footsteps of Bridgland and Brinkman returning for him. Leaving the cairn to join them, Harris became so disoriented that when Bridgland asked him where their camera station and cairn were he was reduced to guessing. Even though he rarely indulged in Wheeler-like bouts of anger, Bridgland voiced his displeasure with Harris. He explained that he had been counting on retracing his steps to the mountain trail according to the station's position. Now he had to locate the marker in the midst of driving snow, no visibility, and dangerous terrain.

By the time they got back to the smooth rock face climbed on the way up, the snow had turned to rain, and the cliff was even more slippery. M.P. climbed down first; then the survey instruments were gently lowered down to him. Brinkman and Harris then stared at each other: who would climb down without the help of the other anchoring the rope at the top? "Jack, on the way up, you were flopping around like a fish," Harris averred before agreeing to go last. All proceeded without further incident, but only once they regained the cover of the tree line could they escape the driving rain. More than a decade later, with a frankness that betrayed a frustration with the day's work that probably would have boiled over as rage in the rest of us, Bridgland recorded his impressions of the end of the day: "Cold, wet and tired, we finally arrived at camp, rather disgusted, but thankful that we had succeeded in completing the work."[14] Bridgland never mentioned his plan to name the peak after Harris but certainly demonstrated his dry sense of humour in doing so.

With the summer of 1919 being a bad one for forest fires in the Rocky Mountains, the crew had to work hard to complete their photo-topographic survey of the Bow River and Clearwater forests. Harris remembers that keeping ahead of the smoke necessitated climbing seventeen peaks in twenty-one days; in several of the survey photographs a massive forest fire can be seen burning in the distance. In 1920, wet weather hampered the survey on the outskirts of the Rockies, a little north of the previous year's work, mostly up the Ram River. This work had been commissioned by the Department of the Interior's Forestry Branch, which sought accurate information about the quality, quantity, and accessibility of the timber limits. The areas surveyed now form part of the Wildland Recreational Area and Siffleur Wilderness Area.

A singularly uneventful year, 1920 was followed by Bridgland's poorest survey season ever. In 1921, he and Harris went out to the lower mainland of British Columbia to survey the area around Pitt Lake, near Vancouver. Plagued by rain and low clouds, they decided to suggest the name of August Peak for one mountain in the area simply because it was the only one they were able to climb all summer.[15] With boats for a change, rather than horses, they spent the finer days packing their tents, provisions, and equipment up the creeks that emptied into the lake, hoping beyond hope that they would have enough hours of respite from the rain to set up camp before a thorough soaking descended on them.

BY THE TIME THEY GOT
BACK TO THE SMOOTH
ROCK FACE CLIMBED
ON THE WAY UP,
THE SNOW HAD
TURNED TO RAIN, AND
THE CLIFF WAS EVEN
MORE SLIPPERY.

Despite poor conditions that summer, the men did manage to accomplish some work. One day, Harris and Claus Nicholson had gone to set up camera stations on a low peak at the northwest end of Pitt Lake. They left camp in a small boat at seven in the morning and rowed to the base of what appeared like a quick climb, but were gone until nine that evening. When they returned that night, Bridgland exhibited no surprise that it had taken them so long: because Pitt Lake is at sea level, even though the peak measures only 5,595 ft (1,705 m), the men had had to ascend the full height, unlike in previous survey areas where they often started climbs from camps that were a few thousand feet above sea level. As the weather was not conducive to work in camp on the evening of Harris' return, he took the opportunity to ask about M.P.'s career. After his long day's hike and perhaps hoping to rival his chief with his day's total, he asked M.P. about his own greatest single-day gain in elevation. The astonishing reply was 7,600 ft (2,316 m), although Bridgland allowed that he covered the height between two in the morning and eleven at night—more than a day except in the twenty-four-hour sense of the word. Even so, it was a formidable elevation gain; still today, experienced mountaineers regard any single-day elevation gain of more than 2,000 m as a notable achievement.

Bridgland shared many stories that evening, including several about the deceptiveness of distances in the mountains. He told Harris of the time he took Alpine Club of Canada guests onto Abbot Glacier at Lake Louise. As the group hiked along the path at the top of Abbot Pass, M.P. asked the climbers how long they thought it would take them to reach the peak. They guessed between twenty and thirty minutes. He responded only by smiling: experience had taught him that it would take another five hours. As for guessing the time of various descents, he admitted that, even with his many years' experience, he was quite taken aback when, after climbing a glacier on Mount Purity (10,331 ft [3,149 m]) in the Selkirks for five arduous hours, the party was able to glissade to the base in only twelve minutes.

✳ *Surveying Kootenay Park, 1922–1923* ✳

IN 1922 AND 1923, the Mapper of Mountains was called back to the majestic peaks of the main ranges, this time to survey parts of Kootenay Park, which had been established only in 1920, and which did not yet have comprehensive maps of its ranges. During these two years

Three photographs of smoke and the results of it,
from the 1919 and 1920 survey seasons.

One (*top*): Photograph by M.P. Bridgland, Bow Forest Survey, B–222–19,
Stn 254, no 222, direction south, 1919.
Two (*bottom*): Photograph by M.P. Bridgland, Bow Forest Survey, B–260–19,
Stn 258, no 260, direction northeast, 1919.

Three: Photograph by M.P. Bridgland, Bow Forest Survey, B–233–1920, Stn 344, no 233, direction south, 1920. Courtesy Library and Archives Canada R214.

Photograph by M.P. Bridgland, Bow Forest Survey, B–434–19, Stn 282, no 434, direction southwest, 1919. This is one of Bridgland's most sublime alpine photographs. It depicts the glacier and three unnamed tarns at the head of Martin Creek, in the Clearwater River watershed, even today a remote district in the northeast corner of Banff National Park. From left to right are mounts Harris (named for Bridgland's assistant, Ley Harris), Willingdon, and Clearwater. Courtesy Library and Archives Canada R214.

he also surveyed north of the park, down the Columbia River valley. Like every other region, Kootenay offered several challenges and adventures. The most unfortunate mishap of the season involved a traverse of Kootenay River. The party had to cross at a point that was especially swift. Trapper Bill Yearling (who spent five years in the area from 1920 to 1925 before selling his 260-acre parcel of land that is now the Nipika Mountain Resort[16]) operated an aerial ferry on the river and the men availed themselves of it, although the prospect of being hauled by old fraying ropes some metres above a rushing torrent did not delight Harris. Last to cross was the survey equipment. Suddenly, the ferry ropes snapped and the packs all fell into the rushing water. They were still attached to one end of the rope and Yearling was able to pull them ashore on his side, but he had to re-rig the ferry before he could send them over to their owners. The day's miseries did not end there: the pack horses were too heavy to transport on the ferry, and so, before crossing the river himself, Bridgland had walked them upstream so that they could swim across without being swept past the point where the men planned to land. Giving them a quick slap on the rump, Bridgland sent them in. Used to this sort of arrangement, they began diligently to work across the fierce pull of the water, with just their ears, noses, and tails visible. Despite Bridgland's precautions, they were all swept farther downstream than anticipated and one horse drowned.

Another incident in Kootenay could easily have been fatal but fortunately proved otherwise. In a valley near Windermere Road, Bridgland and Harris set out one morning with their respective assistants to climb two adjacent peaks. As they photographed and took transit notes, they both noticed but were not overly concerned by the sight of gathering thunderclouds in the distance. Harris' assistant then cried out in amazement. The fringes on his chief's buckskin coat were sticking straight out from his body, making him look like a giant bottlebrush. As the assistant drew nearer, the fringes dropped to their normal position, and then as he moved away again Harris' jacket resumed its comical but alarming appearance. Both men could feel the static electricity in their hair and were developing acute headaches. In the meantime, Bridgland had climbed higher on his peak than his assistant, and was surveying from a camera station on a lonely ledge when he felt someone brush his shoulder rather forcefully. He looked around but no one was there. It didn't take him long to put two and two together, however. He prudently descended from the ledge, found his assistant,

THE FRINGES ON HIS CHIEF'S BUCKSKIN COAT WERE STICKING STRAIGHT OUT FROM HIS BODY, MAKING HIM LOOK LIKE A GIANT BOTTLEBRUSH.

and commanded him to get off the mountain as quickly as safety permitted. Back at camp he and Harris excitedly exchanged descriptions of how they could see fire flashing from their ice-axes every time they hit rock, and how all the metal instruments were singing and buzzing. Mild static electric storms were not unusual in the mountains, and Bridgland experienced them at both high and relatively low elevations, but this particular day produced certainly the most pronounced example, and one from which the men very gladly escaped unharmed.

❋ *Surveys Engineer, 1924–1931* ❋

IN 1924, A REORGANIZATION OF THE DEPARTMENT OF THE INTERIOR changed Bridgland's title to Surveys Engineer, Grade 4, and his annual salary was altered to $3,300. He again worked in the area around the Clearwater Forest Reserve. As his career progressed and the Great War gave way to the prosperous 1920s, the summer months brought increasing numbers of people to the mountains, and he often encountered tourists, packers. outfitters, rangers, Native people, mountaineers, and other surveyors. In the summer of 1924, he met Bert Alford, the newly-commissioned assistant forest ranger for the area, and his wife, Ethelwyn Octavia Doble Alford, who bravely agreed to spend the summer with her husband even though severe rheumatism afflicted her at the age of fifty-two. The Alfords often crossed paths with Bridgland and his crew, and they visited and dined together when opportunity arose. On 19 July, just after the Alfords had replenished their provisions with some fresh venison sold them by Silas Abraham, a Stoney man who lived in the area, Mrs Alford treated Bridgland and his men to a delicious hot meal of pot roast, roast potatoes, spinach, dumplings, rhubarb pie, and tea. This was the best cooking the men had enjoyed since leaving Calgary two months earlier, and they were only too glad to return the favour in August, when the Alfords travelled with M.P. for about a week. Mrs Alford was not feeling well for most of the summer, and she found it quite a relief to let Bridgland's cook prepare the meals while she rested or did washing or other chores for her husband. Likely, her ability and willingness to work despite her rheumatism earned her Bridgland's respect and sympathy. After they parted company, Bridgland sent his cook with two loaves of bread to Mrs Alford at the ranger's cabin where the couple was staying. In addition, M.P. had one of his men chop wood and make the cabin ready for them,

MRS ALFORD TREATED BRIDGLAND AND HIS MEN TO A DELICIOUS HOT MEAL OF POT ROAST, ROAST POTATOES, SPINACH, DUMPLINGS, RHUBARB PIE, AND TEA.

prompting Mrs Alford to note in her diary that "wherever one goes in the world ... you meet with kindness from strangers."[7]

Bridgland had long been the senior employee at the Calgary branch of the Topographic Survey but the reorganization of 1924 put the office in his charge (it moved that year from the old Thomas Block to the Southam Building at 1st Street and 7th Avenue SW). He had a staff of ten men, all of whom helped with the various tasks involved in map-making, whether in the field or at the drafting table. Over the next few years, he supervised and organized surveys in Calgary and the districts around Revelstoke, as well as the valleys of the upper Kootenay and upper Columbia rivers.

In 1924, as well, after a few years of experimentations with aerial photographic techniques, the Department of the Interior decided to adopt this method of survey on a large scale. Nevertheless, the department published *Photographic Surveying*, Bridgland's own authoritative booklet (which, in the year of Deville's death, deliberately retained the title that Deville had used for his book thirty-five years before). It was distributed to survey offices across Canada and was often forwarded to foreign survey agencies in reply to requests for information on the latest developments in photogrammetric techniques. But the use of airplanes was definitely on the horizon.

The 1925 survey of Calgary gave the forty-seven year old his first ride in an airplane and his introduction to aerial surveying, the technique that Surveyor General Deville anticipated as the culmination of phototopographical work. Experiments with this new technique had been refined year by year. Bridgland learned that surveying mountains from an airplane yielded comparatively small-scale photographs because the aircraft had to fly high above ranges to reduce the difference in scale between high and low peaks in the photographs. Although the new technique interested him, he concluded that nothing could rival mountaineering phototopography. His temperament was vital to his preferred methodology, however; the work demanded extreme patience and robust physical fitness and stamina. Would any surveyor armed with Global Positioning Systems be up to such a demanding task today? Surely most men would have greeted aerial photography as a great advancement on methodology.

Content to return to *terra firma*, Bridgland also spent most of the 1925 field season in town, while Ley Harris conducted work between Revelstoke and the Shuswap Lakes. Few mountains in this area exceed

9,000 ft, but because the valleys are low the exhausting season featured long hikes to low peaks through dense brush. Harris and his team managed to establish ninety camera stations but only eight of them stood above 8,000 ft. One might conclude from this fact, and from the gap for 1925 in the chronological list of peaks climbed which M.P. submitted to the *American Alpine Journal* at the end of his career, that the summer's climbs offered no challenge to the alpinist in him. In truth, however, Bridgland was feeling his age. In an aside in a letter to his superiors concerning the priorities for the 1925 field season, he put them on notice that he was looking for less demanding assignments in the future. As the 1920s wore on, he assigned an increasing amount of the phototopographical, higher-elevation work to his assistants while overseeing the work and concentrating on the less demanding stadia traverses. Two full decades of work-related and recreational climbing had begun to take their toll.[18] So it is all the more intriguing that he did not show much interest in aerial photography; it would have given him some relief from the punishing climbs that his preferred method required.

But another reason probably accounts for his disinclination to dive into new realms of work. As early as June 1907, after outlining details of his modest income since the inception of his career, he told the surveyor general that the work he was doing ought to be worth more than he received, "at least the fieldwork. I have been informed that assistants in any position are entitled to $5.00 per day while in the field and would be grateful to you for any information on the subject, and still more so if you will be kind enough to do anything for me." Twenty years later, he was still feeling under-remunerated: "Photographic Mountain work," M.P. told Surveyor General Peters, "is work of a special class, and as such is entitled to special recognition." He outlined the dangers, difficulties, and physical and mental stresses of mountain surveying and of the tedious and strenuous office work, and then asked Peters to "give the staff of the Calgary Office due consideration" when making proposed changes. Later in 1927, he complained to Deputy Minister of the Interior W.W. Cory that, "since Mr. Wheeler left the service of the Department in 1910, I have been the only man in the Branch in charge of this class of work and have made detailed surveys of a larger area by photographic methods than any other man. ... I wish too, to call attention to the importance of this work in connection with the development of the mountain regions. ... The maps

"PHOTOGRAPHIC MOUNTAIN WORK," M.P. TOLD SURVEYOR GENERAL PETERS, "IS WORK OF A SPECIAL CLASS, AND AS SUCH IS ENTITLED TO SPECIAL RECOGNITION."

supplied by our earlier surveys have been of the utmost value in developing and stimulating tourist traffic."[19]

Over the years, the obligation to write innumerable memos, letters, and telegrams to the surveyor general's office fostered in Bridgland an understandable distaste for the administrative side of his job. In the summer of 1925, however, administration took on a humorous dimension during a spot of fieldwork. Frederic Hatheway Peters (1883–1982) became Deville's successor as surveyor general, serving from 1 December 1924 (three months after Deville's death) to 14 January 1948 (retiring the day before Bridgland died). He decided to visit Bridgland in the field to learn for himself what phototopographical surveying was all about. He had worked primarily with the irrigation surveys around Calgary before his promotion to surveyor general but had never climbed mountains. Bridgland invited him to a camp he'd established at Crazy Creek, west of Revelstoke, and near the then-abandoned lumber mill at Taft (see map on page 56). The day dawned clear and bright when M.P. and Ley led the way up to Crazy Creek Lookout where the Forestry Department had a fire station. The surveyors carried the transit, tripod, camera, and negatives so they could do their regular survey work. The three men seemed to be making good progress up the easy trail, but when Bridgland turned to make sure Peters was following he beheld an obviously frightened figure down on all fours, crawling up a path that was an easy stroll, albeit a steep one, for Bridgland and Harris. The surveyors tactfully said nothing, continued to the summit, and took their photographs and transit readings. A few hours later, the men were back in camp with the lookout man as an extra dinner guest, but Peters was too exhausted to visit and chat. He said that he felt it was a bad idea to eat when he was so tired, but the surveyors suspected that he was simply too tired to lift a fork to his mouth. Another flatlands surveyor bit Rocky Mountain dust that day.

The following year, Bridgland again worked on surveys in the areas of Calgary and Revelstoke. As well, he introduced himself to the flatlands of the prairies, following the Bow River eastward as far as Gleichen and Drumheller. In this year, his younger son Edgar was recruited to help out in the camps. Usually, he would hold the rod while his dad sighted. Edgar assisted only a few times at mountain survey camps but occasionally worked with his father on the prairie camps around Edmonton and Calgary, where they would use a plane table and transit.

THE THREE MEN SEEMED TO BE MAKING GOOD PROGRESS UP THE EASY TRAIL, BUT WHEN BRIDGLAND TURNED TO MAKE SURE PETERS WAS FOLLOWING HE BEHELD AN OBVIOUSLY FRIGHTENED FIGURE DOWN ON ALL FOURS, CRAWLING UP A PATH THAT WAS AN EASY STROLL, ALBEIT A STEEP ONE, FOR BRIDGLAND AND HARRIS.

In 1927, while working on the southern end of Jasper and the northern end of Banff parks, Bridgland met with the second fatality in his career when one of his assistants, name regrettably unknown, slipped and fell to his death while climbing Mt Wilson (10,696 ft [3,260 m]). Joan Robson, a noted botanist from Jasper who was then hiking from Jasper to Lake Louise with Jack Brewster and a number of other guests, remembers meeting Bridgland and hearing of the incident:

> July 10 was rainy, but we forded [Nigel] creek, crossed the site of the present Jasper-Banff Highway, and climbed a hill across the valley from camp, where we got our first view of the Saskatchewan Glacier, with its prominent medial moraine. The usual glacial downdraft was present, so we soon cooled off and headed for camp. Mr. Bridgeland [*sic*], Dominion Land Surveyor, had visited us in camp the night before and we were stunned to meet one of his men when we reached camp, as he was looking for Mr. Bridgeland to report that one of his surveyors had lost his life on Mt Wilson in the afternoon.[20]

From 1927 to 1929, Bridgland supervised three different surveys, one in the Revelstoke area, one in Calgary, and one in Jasper Park, but only in the last did he do the surveying himself. During his last mountain field season, he was especially happy to have the assistance of Pete Benner and Pete Benoit, two reliable and efficient packers who had worked with him many times previously. The fact that they came back year after year indicates fairly reliably Bridgland's congenial nature and confident management of a survey team, for the job of a packer is very physically demanding and offers few rewards; few men made a habit of taking on the position for more than a year or two.

In 1930, while still managing the Calgary office and overseeing the phototopographical surveys conducted from there, M.P. undertook fieldwork, mostly township division surveys, in the Peace River district of Alberta and British Columbia, which was burgeoning with new settlement and agriculture. This survey used a rod and a transit rather than a camera. While it did not interest him very much, it was all the work he could handle in addition to office duties and his work as a councillor with the Association of Dominion Land Surveyors and as a committee representative with the ADLS and the Alberta Land Surveyors' Association.

✳ *Forced Retirement* ✳

IN 1931, survey work on the boundary of Banff National Park was suddenly discontinued as a result of a reorganization of federal and provincial land agencies. After natural resource rights were granted to the western provincial governments, Prime Minister R.B. Bennett, whose Progressive Conservatives had swept into power in 1930, drastically cut funding to the federal surveys. Many Albertans were pleased by Premier Brownlee's acquisition of provincial control over resources, but Bridgland proved to be one of the losers in this transfer. The days of the department were nearing an end. Not just Brownlee but Prairie politicians generally had gradually grown restless with the federal government's continued control of western Dominion Lands and resources in contravention of the assignment of them to provinces by the *British North American Act*. When finally in 1930 control of them transferred to the provincial governments, an end came to the sixty-year existence of Dominion Lands except in national parks, Indian reserves, and the regions of Canada that did not have provincial status. A massive reorganization of provincial and federal government administrations followed. The fate that was meted out to Bridgland was hardly isolated.

The other reason that prompted the curtailment of photopographic surveying has already been suggested. Ever since the first maps of Canada were made, the development of new survey instruments and techniques fundamentally affected the nature of the geographic information recorded and the maps produced. The introduction of photographic survey techniques was the most significant innovation in Canadian, and indeed, global mapping practices. Once the basic principles of photogrammetry had been proven highly efficient in mountain surveys, the progression to aerial photogrammetry seemed not only logical but inevitable. Even before the advent of the airplane, it was predicted that survey photographers would lose their jobs as soon as hot air balloon technology improved and allowed vast landscapes to be captured in a single day's work. In the event, at least in Canada balloons never were used for surveying, mostly because early pilots had neither the ability to determine their exact altitude nor the technology to ensure consistent altitude for every photograph. Airplanes, however, offered both stability and altimeters. Although aerial photography cannot be completed without field surveyors to establish control points, the relative efficiency of photography from airplanes and, later, helicopters led

ONCE THE BASIC PRINCIPLES OF PHOTOGRAMMETRY HAD BEEN PROVEN HIGHLY EFFICIENT IN MOUNTAIN SURVEYS, THE PROGRESSION TO AERIAL PHOTOGRAM- METRY SEEMED NOT ONLY LOGICAL BUT INEVITABLE.

to the drastic reduction in the numbers of mountaineer-surveyors.[21] Aerial photography would indeed eclipse oblique-angle mountain-top photography but not before Bridgland had perfected his method and issued his remarkable maps of the Rockies.

On 12 May M.P. received unexpected news from Surveyor General Peters: after thirty years of service, he was dismissed. Giving him just over a week's notice, Peters' bluntly worded letter stated that "the reduction in the amount available for the continuing work of this Department makes it necessary to terminate your services from and after the 20th instant."[22] Further correspondence divulged that the position of Surveys Engineer, Grade 4, had been abolished. What a blow! M.P.'s job just disappeared out from under him, like scree on a treacherous alpine slope. He spent much of the rest of May scrambling, trying to gain an assurance from the department that he would be awarded the six months' retirement leave owed him, but the department would grant him only three.

Despite the abrupt tone of the letter, the department kept him working until December and beyond, although it recorded his official retirement date as 31 August. He was employed as a daily-paid surveyor on the Banff National Park boundary. But, as the department stipulated, that no time was to be allowed after the work was done for the compilation of notes, Bridgland handed over his negatives without completing the map for which they were taken. He was only fifty-one years old but at the end of his days as a mapper of mountains. His career came to an end for good when he baulked at what he considered an insult. Writing to Peters in November 1931 after his position had been terminated and yet he was offered contract work in Riding Mountain National Park, Manitoba, he said that he did "not care to undertake the work in the Riding Mts. as the work, responsibility, and exposure to winter conditions are out of all proportion to the remuneration. As you know, I have always considered it my place to accept any work assigned to me while an employee of the Department. Now that I have been let out the situation is different. If I am asked to take work under the most disagreeable conditions knowing that I am to be let out as soon as the work is completed, some special consideration should be given." He was beyond diplomacy. Sadly, he left his employ dispirited and feeling slighted.[23]

Such a change of fortune left the dedicated surveyor in shock, from which he never fully recovered. As his friend Charlie Sissons noted,

HE WAS ONLY FIFTY-ONE YEARS OLD BUT AT THE END OF HIS DAYS AS A MAPPER OF MOUNTAINS.

"He rarely permitted himself to speak of this event. But in 1939, when I was studying his map of the [Mt] Coleman region [in northeastern Banff National Park] and asked for information as to what lay to the northeast, he told me of his regret at not being permitted to finish the mapping of this area."[24] Edgar Bridgland, M.P.'s younger son, remembers his father's thinking that he had fallen victim in a struggle for dominance between the Geological Survey and the Topographical Survey, with the former prevailing. No doubt, however, the Depression also played a considerable role in the slashing of both agencies' budgets. All in all, M.P. had gone about his work under circumstances that most of us would not countenance from an employer today. The terms on which the DLS operated were strict to the point of cruelty, and the onus seemed always to rest with the surveyor in the field to account for each and every decision taken and expense incurred. Whether it was Peters or Deville in the surveyor general's office, a person of Bridgland's dedicated achievement deserved better treatment. Consider a letter he had to write on 27 September 1915, during the summer's survey in Jasper. It pertained to the previous summer's and winter's expenses, some of which were still being questioned or compensation for which had been denied:

> Sir:
> I am in receipt of your letter of the 21st inst. Enclosing a copy of balance sheet for 1914.
> As I have no copy of my accounts here [in Jasper Park], I am not in a position to express an opinion relative to vouchers 1, 2, 4, 7, 8, 9, & 11 which have been disallowed. In connection with the Crowsnest map I had to get water colors to use in place of ink and I also hunted all over Calgary to get tracing linen suitable, and wide enough for the plan. Items of this nature I think, should be allowed, but if your decision is otherwise, I have nothing further to say. Running an office all winter and preparing a large plan is rather different from ordinary work.
> Regarding the deduction of $9.00 for one "day off" I am certain that you do not understand the circumstances. Evidently my diary has been examined by a clerk who knows nothing of mountains, and who has not even read the diary carefully. If I remember correctly my diary shows that we worked for the seven days preceeding [*sic*] and for the nine days following the "day

ALL IN ALL, M.P. HAD
GONE ABOUT HIS WORK
UNDER CIRCUMSTANCES
THAT MOST OF
US WOULD NOT
COUNTENANCE FROM
AN EMPLOYER TODAY.

off"[.] During this period we did the hardest work and made the best progress of the whole season. During the two months spent on triangulation this was the only day lost except through bad weather, and I consider that that day was earned.

As you know, it is practically impossible to observe Sunday on this work, and instead I have always given the boys a day off when convenient. In the majority of cases, this has been a day when we could not work and the Department has been that much ahead. Occasionally I have considered it advisable to take a fine day off as in the case under discussion. This I have always shown in my diary and it has never before been questioned. I do not consider it just that this time should be charged against me, neither do I consider it proper that I should have to doctor dates in my diary in order to make it appear that we worked all week and took Sunday off.[25]

Thus coldly does the letter come to a halt; nor does the typescript include a closing salutation, just "M.P. Bridgland." It seems a monstrous injustice to follow a conscientious man through the mountains, in war time no less, hounding him about every little cent, all of it legitimately spent on his work, none of it wasted on extravagances. What an even temper must one have had to suffer the thoughtless, inhuman bureaucracy out in Ottawa. Had A.O. Wheeler been right to throw over the DLS when it refused him and M.P. leave in 1911 to run the ACC summer camps? Sounding his most exasperated, M.P. must have wondered. Perhaps Peters' letter in May 1931 struck him with less surprise than we might imagine; in some ways it ran true to form. He might have thought delightfully back to that summer when he and Harris had spotted Peters crawling along on the trail. But, no, that sort of resentment doesn't seem characteristic of him.

One benefit that accrued from Bridgland's forced retirement was the time it afforded him for his family. In 1931, before the worst of the Depression hit, the entire family travelled together from Calgary to Toronto, exploring many stops along the way. During both his career and his retirement, M.P. dedicated himself to introducing his family to much of the country. Aside from short trips to the mountains, or to Banff for the annual Indian Days, the family visited Vancouver, Victoria, and Cardston, the latter in about 1921, just before the Mormon temple was completed, when Edgar was six years old.

While out east in September, Bridgland was honoured by his associates at an evening in Ottawa hosted by the Association of Dominion Land Surveyors. Although he was always modest about his accomplishments, he had many of which he could be proud, including the meticulous accuracy of his maps, the professional and aesthetically-pleasing quality of his photographs, and the ascent of literally hundreds of peaks in the Rockies and Selkirks. The Alpine Club of Canada recognized his contributions as a pioneer mountaineer three years later, when at the annual camp (held that year at Chrome Lake, in Jasper National Park south of Tonquin Valley) he was awarded the Silver Rope for Leadership. This distinction was bestowed upon those who demonstrated excellence in alpine guiding and a dedication to the instruction of other club members.

✳ *Bridgland's Later Life, 1931–1948* ✳

FOR A FEW YEARS AFTER RETIRING FROM SURVEY WORK, Bridgland continued to find much to do in the mountains. He remained active in the Alpine Club of Canada and often accompanied friends and visiting family on climbs and camps. Margaret Hess remembers her Uncle Morris as "self-contained, but also the most loyal friend you could hope for."[26] Among his closest friends were fellow Calgarians S.R. Vallance, a famous lawyer, and L.C. Wilson, a bookstore owner, both of whom shared Bridgland's love of photography and mountain climbing, and with whom he took several climbing trips in the Rockies. As well, he enjoyed the friendship of Edward, Ernest, and Walter Feuz, the best-known Swiss guides ever to have worked in the Banff area. Indeed, Bridgland made the acquaintance of innumerable alpinists through his association with the ACC, and guided several people who later went on to great fame, if not fortune. Among these were the Reverend Charles William Gordon, who grew famous under his pen name, Ralph Connor; Edward Oliver Wheeler, who made an attempt on Mt Everest in 1921 with Sir George Leigh Mallory, and who was to become surveyor general of India; and, Sir Sandford Fleming, the famous engineer who had also been Bridgland's much more senior neighbour when M.P. grew up on the farm at Fairbank.

Aside from his many surveying and guiding colleagues, M.P. was also a friend of arguably the most famous professional photographers of the Rocky Mountains, Byron Harmon and Harry Pollard. While the

INDEED, BRIDGLAND MADE THE ACQUAINTANCE OF INNUMERABLE ALPINISTS THROUGH HIS ASSOCIATION WITH THE ACC, AND GUIDED SEVERAL PEOPLE WHO LATER WENT ON TO GREAT FAME, IF NOT FORTUNE.

former enjoyed great success in publishing photos that interested tourists in and around Banff, including pictures from several ACC camps and excursions, the latter was renowned for the many photos he took on behalf of the CPR.[27] Surely, these photographers were impressed with the professional quality of Bridgland's work. In 1920, a view he captured of Lake McArthur won the Alpine Club of Canada's annual photo contest. His dedication to photography is nowhere more evident than in his habit of spending hours patiently hand-colouring the black-and-white lantern slides that he created for his talks, lectures, and slide shows from the many trips he took with friends and family in the mountains and elsewhere.

The longest car tour Bridgland ever made was with Edgar in the summers of 1935 and 1936, throughout the midwestern and western United States and western Canada. On this marathon tour, they visited every Canadian and American national park they could reach, and were especially impressed with the awesome beauty of Mesa Verde National and Arches National parks in the American desert, and with the paucity of visitors to the area: as Edgar recalls, they drove once for an entire day and saw only one other car on the road. As well during these last few years in Calgary, Mary and M.P. kept up their interest in religion by occasionally meeting with a Bible study group that would convene at participants' homes. Such groups were common at the time, the most famous being the Prophetic Bible Institute, also a Methodist group, which was established by his contemporary, William "Bible Bill" Aberhart (1878–1943), who served as premier of Alberta from 1935 until his death. While Aberhart's gospel aligned with Bridgland's, his Social Credit policies, including his scheme to print Alberta's own money, were contributing factors in Bridgland's decision in 1936 to leave Alberta. Other factors weighed in the decision to move: Edgar and Charles were going to be living in the east, and, with the Depression nearing its deepest, he felt he could offer more advantages to Edgar in Toronto, near their relatives. Charles was by this time a graduate of the University of Toronto's electrical engineering program, and Edgar had just completed his first year at Victoria University, University of Toronto, in engineering and aeronautical studies.[28] In the West, the Bridglands' only relatives were the Hess family, but in the East over thirty people regularly attended the New Year's Eve and Labour Day family reunions on the farm at Fairbank. That said, Edgar remembers that his father "really had nothing to do but putter around on the farms with his brothers—that's all."

ON THIS MARATHON TOUR, THEY VISITED EVERY CANADIAN AND AMERICAN NATIONAL PARK THEY COULD REACH, AND WERE ESPECIALLY IMPRESSED WITH THE AWESOME BEAUTY OF MESA VERDE NATIONAL AND ARCHES NATIONAL PARKS IN THE AMERICAN DESERT, AND WITH THE PAUCITY OF VISITORS TO THE AREA.

If one were to go to Toronto today and stand at the corner of Dufferin Street and Eglinton Avenue, looking north, it would be difficult to imagine that the area was, just over sixty years ago, still a prosperous farming community. The Bridgland and Parsons families sold off all their farm properties by 1955, and the land was developed into a residential area shortly thereafter. Yet, several landmarks point to the rural history and to the Bridgland family's history there. Fairbank United Church, just a few blocks north of Eglinton, at Dufferin and Wingold Avenue, stands one of the few reminders of the area's old name and its Methodist roots. In fact, this church marks the southern boundary of what used to be the Parsons family's 200-acre farm. A little farther north, the C.B. Parsons Middle School marks the location of fifty more acres of land owned by the Parsons. A little farther north again, Bridgeland Avenue recollects the place where there once was a rocky laneway running through the middle of the Bridgland property, leading to the farmhouse and outbuildings. In 1948 or 1949 the city added an *e* to the avenue's name, thereby misspelling it. M.P. did not fare well by bureaucracies.

When Bridgland retired to Toronto at the age of fifty-eight he did little in the way of regular physical activity, but he looked after himself, neither smoked nor drank, and saw every project he took on to a thorough conclusion. Working long hours in his old captain's chair, he spent months on the meticulous process of colouring the hundreds of lantern slides he had made from his photographs. Library and Archives Canada hold a collection of 715 of these; it comprises both personal views and a selection of Department of Interior photographs of flora and fauna, which he would use to illustrate his lectures on the mountains to friends, family, members of the Alpine Club of Canada, the local YMCA, and meetings of Boy Scouts or Girl Guides. This amounted to one link to his career. Another came to his son's attention in England in 1938. At Victoria and Alberta Museum in London, Edgar saw a scale model exhibit of the mountains based on his father's work. Edgar had no idea who mounted it and the next time he visited the museum he not only could not locate it but could not ascertain where it had gone. Had M.P. worked on this model himself? Perhaps time will tell. For the most part, however, it seems that the move from Alberta to Ontario had essentially cut him off from the geography in which all his abilities and hobbies found their widest expression. Edgar believes his father may have lived longer had he not left the mountains

WORKING LONG HOURS IN HIS OLD CAPTAIN'S CHAIR, HE SPENT MONTHS ON THE METICULOUS PROCESS OF COLOURING THE HUNDREDS OF LANTERN SLIDES HE HAD MADE FROM HIS PHOTOGRAPHS.

Three of the lantern slide photographs made by M.P. Bridgland in retirement:
a bear being fed by hand, Athabasca Gorge, and a marmot on The Whistlers.

Photograph by M.P. Bridgland of the Ramparts and Amethyst Lake from Mt Majestic, Stn 16, no 132, direction southsouthwest, 1915. Courtesy Jasper National Park.

he so dearly loved. An obituary written by Sissons notes that, although "he began and ended his life in the East, his heart was in the West, and particularly in the Rockies, in the surveying, photographing and mapping on which his active life was spent."[29]

Morris and Mary had been married forty years when he died at age sixty-nine of leukemia at home, 64 Snowdon Avenue, Toronto, 15 January 1948. Born the same year, Mary lived another quarter-century after his death.

❋ Mapper of Mountains ❋

OVER THE COURSE OF HIS CAREER, Bridgland completed photographic surveys in many mountain areas, most of which were (or became) designated park lands or forest reserves.[30] These included large blocks of Jasper and Glacier parks, the Columbia and Kootenay river valleys, the Bow River, Clearwater, and Crowsnest forest reserves, and the Revelstoke district in British Columbia. For the most part, these lands were adjacent to the mountains' main thoroughfares, whether a railway line, one of the main rivers, or a road. As the years went by and more roads were built as more areas were mapped, roads and railways became the arteries through which tourism flowed in the maze of the mountains. His utmost desire was to make the mountains as accessible as possible to as many as possible, but, although the roads and railways facilitated access, they also destroyed parts of the natural ecosystems. He knew that the railway had been built to unite the country but was astute enough to notice that its construction in fact divided animal habitats. Although he saw this aspect of the railways and roads as a repercussion, he was confident that park wardens would implement management policies to restore and protect animal populations. Like most people of his generation, M.P. did not consider irreversible the effects of pollution from vehicles and holidayers.

M.P. (*upper left*) with his sister and brothers, c.1945.
Courtesy Edgar Bridgland.

The period from about 1870 to 1930 was a time of great excitement and discovery in the Rocky Mountains; entire libraries exist of the personal accounts of professional explorers, government geologists, casual tourists, and avid mountaineers who helped develop the rich lore of the Canadian Rocky Mountains.[31] In 1909, when Charles Walcott, the famous geologist and secretary of the Smithsonian Institution, publicized the discovery of fossil beds at the Burgess Shale in Yoho Park (just one range over from where Bridgland surveyed that summer), he excited public interest in the 515-million-year-old fossils and in the scientific value of the mountains generally. This widespread excitement in turn boosted the federal government's confidence that economic profit could be derived from the mountains' resources. As well, just months before the First World War broke out in Europe, the discovery of oil in Turner Valley, southwest of Calgary, started Canada's first oil boom. Canadians began to recognize the resource potential of the western mountains. Despite the shortage of funds and able bodies lost to war, the Department of the Interior ordered surveys throughout the Rockies, particularly in especially beautiful or valuable districts. In part because of Bridgland's surveys of the area and his classifications of the land, more of the Selkirk Range, between Golden and Revelstoke,

British Columbia, than had been embraced by the establishment of Glacier Park in 1886 was set aside as Mount Revelstoke Park in 1914.

Charles Sissons called Bridgland "one of the greatest of Canadian mountaineers,"[32] and Edward Oliver Wheeler thought that he had "rendered all mountaineers incalculable service not only in mountaineering as such but in making possible mountain travel based on sound and accurate maps."[33] Indeed, to this day the climbing routes and hiking trails he pioneered and documented, the resources he mapped, the cairns and huts he helped to build all continue to make remote parts of the mountains identifiable and accessible to every type of mountain visitor, from alpinist to casual tourist and from investor to lumberjack. On almost every survey, Bridgland completed dozens of ascents, counting at least fifty-five firsts over the course of his career.[34] To this day, hundreds of mountain peaks and dozens of rivers, valleys, and glaciers retain the names Bridgland gave them. Aside from these records of his legacy as a surveyor, one mountain bears his own name: Mt Bridgland (9,613 ft [2,930 m]), in the Victoria Cross Range in Jasper National Park, was named for him by Frank Sissons in 1923. As well, Bridgland Peak (9,850 ft [3.002 m]), Bridgland Pass, and Bridgland Glacier honour his name in the Selkirk Range about twenty km north of Mount Revelstoke National Park.

It takes nothing away from James Joseph McArthur and Arthur Oliver Wheeler to state that no one deserves to be called Mapper of Mountains more than Morrison Parsons Bridgland. With his death in 1948, a mere ninety years after the hardly noticed death of David Thompson, he merits comparison with that early transcontinental cartographer in mapping achievement, and, if one credits vertical as well as horizontal distances covered, perhaps even in territory charted. One obituary notes, "Mr. Bridgland gave practically his whole active field service to this class [phototopographic] of surveying and became recognized as a world authority in Phototopographic Surveying."[35] Apart from his different method—astronomical—of surveying, no more or less has been claimed for Thompson. As we get set to celebrate the bicentennial of Thompson in the West, we also are able to take great interest in the near century since Bridgland's work. And Bridgland's photographs, more than Thompson's maps, have left us a legacy that we are only now beginning to appreciate and exploit in order to learn how the influences exerted on non-human nature over time can be measured, analyzed, and understood. ❋

CHARLES SISSONS CALLED BRIDGLAND "ONE OF THE GREATEST OF CANADIAN MOUNTAINEERS."

✳ Notes ✳

1   The book's preface includes the following statement: "The topographical part of the Guide was written by him [Bridgland] and the historical notes by R. Douglas, Secretary of the Geographic Board. The illustrations, which are mostly from the survey photographs, were selected and arranged, and the book edited by E. Deville, Surveyor General."

2   A travel-pouch form of Bridgland's 1917 *Map of the Central Part of Jasper Park, Alberta* was separately for sale in 1923.

3   *Description of & Guide to Jasper Park*, [12]. The poem appeared first in the *Montreal Gazette*, 2 July 1914, and in seven other newspapers by the end of August (Richard Lancelyn Green, "Biographical Note," in Doyle, *Western Wanderings*, ed. Roden and Roden, 8). It must have remained well known, for the Sessional Papers of the Dominion printed it again in 1916. According to Green, Doyle collected it first in his volume, *The Guards Came Through, and Other Poems* (London: John Murray, 1919).

4   On the era's railway pamphlets and other textual representations of Jasper Park, see Gabrielle Zezulka-Mailloux, *The Nature of the Problem: Wilderness Paradoxes in Jasper National Park*, PhD. Diss., University of Alberta, 2005.

5   *Description of & Guide to Jasper Park*, 42–45. The description by Grant is found in *Ocean to Ocean*, 235–36.

6   J.B. Harkin to R.A. Gibson, 13 July 1920, LAC, RG 84, vol 146, file J113-200/2; qtd in Great Plains Research Consultants, "Jasper," 254.

7   A.O.Wheeler and Elizabeth Parker, *The Selkirk Mountains: A Guide for Mountain Climbers and Pilgrims* (Winnipeg: Stovel, 1912).

8   A.O. Wheeler, M.P. Bridgland, and A.J. Campbell, "The Application of Photography," 79.

9   *Description of & Guide to Jasper Park*, 30, 31.

10 One qualification is necessary: in Bridgland's day and for quite some time after, very little thought was given in promotional literature about the mountain parks to visitation during any but the late-spring-to-early-fall period. There is no indication that Bridgland anticipated wintertime visitors. He himself spent little work-related time in the mountains in that season.

    Partial information about the collaboration of Bridgland, Douglas, and Deville is available in Bridgland, "Surveys, Mapping and Remote Sensing," LAC, RG 88, vol 353, file 15756.

11 Lizzie Rummel conducted and tape-recorded two interviews with Harris, the first 4 Sept. 1972, and the second, in Harris' home in Canmore, Alberta, 16 Mar. 1976. (Elizabeth Rummel Fonds, Whyte Museum of the Canadian Rockies, M28/S9/V554; cassette copies held by Association of Alberta Land Surveyors, Edmonton.) A third interview, which provided some of the anecdotes related here, was recorded by R. Dolgoy of the Provincial Archives of Alberta, 13 Feb. 1978, when Harris was eighty-seven years old. (R. Dolgoy, Phonotape, Provincial Museum of Alberta, 78.65/17.)

For more on Lizzie Rummel, who was born into turn-of-the-century German aristocracy, see Ruth Oltmann, *Lizzie Rummel: Baroness of the Canadian Rockies* (Exshaw, Alberta: Ribbon Creek Publishing, 1983).

12 Archives of the Surveyor General of Canada Lands, Legal Surveys Division, Earth Sciences Sector, Department of Natural Resources Canada.

13 Harris states that he suggested the last name to Bridgland after noting that the ridges they had occupied for their camera stations featured peaks that ended in right angles, reminiscent of the shape of dormer windows in a pitched roof, an architectural style that an eastern Canadian would know well.

Bridgland's map of this section of the country includes Ya-Ha-Tinda (Mountain Prairie) Ranch, on the upper Red Deer River about three hours' drive from Calgary and outside the eastern boundary of Banff National Park. Bridgland and Harris' stadia traverse of the area of the ranch has permitted Parks Canada staff to locate their original survey monuments and posts, along with the locations of buildings and trails that existed in 1918–19 (information courtesy Rob Watt, senior park warden, Waterton Lakes National Park). Ya-Ha-Tinda Ranch once fell within the borders of Banff National Park but was excluded for the last time in 1930. It remains under parks protection and management, however. In Bridgland's day, when it served as both the warden district headquarters for Banff Park and the over-wintering pasture for the horses used in backcountry patrols in all the mountain parks, it was known as Brewster Ranch. See Bridgland, "Some Trails between Banff and Nordegg," *Canadian Alpine Journal* 18 (1930): 51.

14 M.P. Bridgland, "Some Trails," 54.

15 Apparently, names are not adopted as official that have their source in irony, ruefulness, or flippancy for no mountain in British Columbia today bears the name August.

16 See "The Nipika Story," http://www.nipika.com/html/about/our_vision.asp

17 Ethelwyn Octavia Doble Alford, "With the Forest Rangers in the Rockies: 2 July 1924-5 October 1924," Unpublished Diary (1924), 13. Photocopy of MS held by University of Alberta Library, Edmonton.

18 "I wish to say that I have been in the mountains since 1902, and while I have no complaints to make at present, I feel that I cannot continue work indefinitely and I would like to discuss the possibility of being given less strenuous work in the future." M.P. Bridgland to Director, Topographical Survey, Feb. (?) 1925, LAC, RG 88, vol 156, file 18522 (Rocky Mountain Forest Reserve Survey ... 1924–1930).

19 M.P. Bridgland to É.-G. Deville, 11 June 1907; Bridgland to F.H. Peters, 4 Oct. 1927; Bridgland to W.W. Cory, 30 Dec. 1927, LAC, RG 88, vol 167, file 1996.

20 Joan Robson, "The Glacier Trail: Jasper to Lake Louise, July 3 to 23, 1927," unpublished paper, 1952, courtesy Tom Peterson. Robson's account does not include the name of the assistant who died. Bridgland's report for the 1927 season, which doubtless identified him, is not extant.

21 The several million photographs taken for the aerial survey of the country and now housed in the National Air Photo Library in Ottawa constitute a more or less complete pictorial record of Canada.

Airplanes saw other park-related applications before Bridgland's career ended. His season in Jasper came only five years before an airplane first landed in the park. Fire-spotting from airplanes was introduced on an irregular basis in the 1920s because, after his exposure to it during service in the First World War, park superintendent Rogers was smitten with air travel. Photography to promote tourism and tourism itself occurred by air. In the summer of 1932, just after Bridgland's forced retirement, a tour operator ran a daily float-plane schedule between Lac Beauvert/Lake Edith and Lake Minnewanka in Banff. The golf course at Jasper Park Lodge was contemplated as an airfield by which to deliver tourists directly to the lodge, but the idea was abandoned, only however after it was determined that none of the fairways was deemed sufficiently long to serve as a runway (Great Plains Research Consultants, "Jasper," 30–35).

22 F.H. Peters to M.P. Bridgland, 12 May 1931, Department of the Interior, LAC, RG 88, vol 167, file 1996.

23 M.P. Bridgland to F.H. Peters, 14 Nov. 1931, Department of the Interior, LAC, RG 88, vol 167, file 1996.

24 In this same discussion Bridgland identified the upper Brazeau River as his favourite district of the mountains (Sissons, 219).

25 M.P. Bridgland to É.-G. Deville, Surveyor General, 27 Sept. 1915, LAC, RG 88, vol. 142, file 13913. According to Bridgland's diary for 1914, the disputed "day off" appears to have been Wednesday 29 July, when he was with his triangulation crew, and had moved into Banff from Fatigué Mtn, en route to Forty Mile Creek (LAC, RG 88, DLS, Field Book no. 14071).

26 Margaret Hess, interview with Gabrielle Zezulka-Mailloux, University of Calgary, 25 Feb. 2002.

27 Both Brock V. Silversides, *Waiting for the Light: Early Mountain Photography in British Columbia and Alberta, 1865-1939* (Saskatoon: Fifth House, 1995), and Edward Cavell and Jon Whyte, *Rocky Mountain Madness: A Bittersweet Romance* (Banff: Altitude Publishing, 1982) offer interesting pictorial histories of the Rockies, and feature many images created by Harmon and Pollard, respectively. See also, E.J. Hart, *The Selling of Canada: The CPR and the Beginning of Canadian Tourism* (Banff: Altitude Publishing, 1983).

28 As noted earlier, Edgar's physics teacher turned out to be none other than Thomas Richardson Loudon (1883–1968), the son of William James Loudon, the professor who had taught M.P. and recommended the DLS summer survey assistantship to him more than three decades earlier.

29 Sissons, "In Memoriam," 218.

30 See page 56–57 for a list of the areas covered by Bridgland's surveys.

31 The Margaret B. Hess Special Collection at the University of Calgary boasts extensive holdings of Rocky Mountains materials, as do The Whyte Museum of the Canadian Rockies, Banff; the Parks Canada libraries in each of the national parks; the Glenbow Museum, Calgary; and the Cameron and Canadian Circumpolar libraries at the University of Alberta, Edmonton.

32 Sissons, *Memoirs*, 167.

33 E.O. Wheeler, "In Memoriam," 222.

34 See appendices for a list of Bridgland's first ascents and selected other climbs.

35 "Morrison Parsons Bridgland," *The Canadian Surveyor* 9:7 (Jan. 1948): 21.

"IF MY ESTIMATION OF BRIDGLAND'S MOUNTAINEERING WAS HIGH BEFORE, IT HAS REACHED EXCELSIOR HEIGHTS TODAY. THIS IS A FORMIDABLE MOUNTAIN. THE WEATHER IS GOOD, BUT FOR HOW LONG THIS IS DIFFICULT TO ESTIMATE."

BRIDGLAND'S LEGACY

CULTURE, ECOLOGY, AND RESTORATION

IN JASPER NATIONAL PARK, AND THE ROCKY MOUNTAIN

REPEAT PHOTOGRAPHY PROJECT

✳

Tracking Ecological and Cultural Changes by Photographs

Thanks to Deville's introduction of phototopography and the perfecting of it by the likes of McArthur, Wheeler, and Bridgland, Canada could lay claim to an impressive inventory of mountain ranges, their geographical features, and some indication of their resources. Dominion Land Surveyors thereby compiled an enduring visual legacy of both the Canadian Cordillera and the process by which it was mapped. Nearly a century later, this vast collection of images is coming to light again, this time to study ecological and cultural changes in the mountains during one of the most tumultuous periods since the glaciers receded more than ten thousand years ago. Today we are realizing what possibilities for comparison over time this vast visual archive has left us. In light of this realization, and although one must keep in mind that for a Dominion Land Surveyor charged with producing maps, photographic surveys were, like triangulation surveys, a means to an end, one legacy of the foremost Mapper of Mountains is rounding into focus.

THE ALPINE CLUB OF CANADA WAS INSTRUMENTAL IN ESTABLISHING IN THE MINDS OF ITS NATIONAL AND INTERNATIONAL READERS THE IMPORTANCE OF PHOTOGRAPHIC COLLECTIONS OF THE ROCKY MOUNTAINS.

Deville and his surveyors appreciated even in the early decades of the century how visual records could assist studies of land use and evolution, as well as efforts to preserve wilderness. And although we tend to think of wilderness preservation as a late-twentieth century development, in fact, it arose as a concerted movement in Canada during the last half of Bridgland's career.[1] In 1901, when Arthur Wheeler started surveying the Selkirks, he began a systematic photographic study of the rate of recession of several glaciers. Joining Wheeler a year later, M.P. quickly grasped the ramifications of such work. Over the course of their several seasons together, they surveyed and measured the limits of the glacial tongues or toes that they encountered. They observed through their photographs the annual recession or advancement of the ice, and in both word and image they published their findings in the annual volume of the *Canadian Alpine Journal*.

The Alpine Club of Canada was instrumental in establishing in the minds of its national and international readers the importance of photographic collections of the Rocky Mountains. In 1918, several members including Bridgland organized a Photographic Committee, which issued the following declaration:

The value of photographic records cannot be overestimated, and increases as they become older. It devolves upon every member of the Club who uses a camera in the mountains to see that nothing which may be of interest escapes him and that all such records, no matter how trivial they may seem now, are placed at the disposal of the Club for future reference.

Its goals were to collect, organize, and preserve photographs in order to promote the art of photography, educate and entertain members and the general public, and provide a "footprint" for future reference and research. In their first annual report, the committee suggested a number of practical uses for its collection, such as helping alpinists pioneer new climbing routes, and illustrating articles in the *Journal* of interest to both scientists and tourists. The report also noted that "[q]uestions of scientific importance will also be referred to the collection, such as the disappearance of lakes, changes in the outline of glaciers, etc."[3]

Such topics absorbed Bridgland's interest and attention, which reached beyond the study of glaciers; he was fascinated by the concept of landscape change. In 1917 he noted in his *Description of & Guide to Jasper Park* that

"ROCKS MOVE," HE WOULD SAY. "MOUNTAINS CHANGE, EVEN IN JUST A FEW YEARS."

> like other glacier streams, Athabaska river carries a great deal of sediment and wherever the current becomes sluggish some of this is deposited. The wide expansions at Jasper lake and Brulé [*sic*] lake form excellent settling basins for this sediment and it is probable that in a few years these lakes will become entirely silted up. The large sand bars visible in low water near the upper end of these lakes indicate how rapidly the change is taking place.[4]

Ley Harris recalled that rock slides and other signs of large shifts in terrain also fascinated his colleague. After surveying in 1913 and 1914 areas that included Coleman and Frank, Alberta, Bridgland often admonished his crews to exercise vigilance in the mountains, regardless of how familiar they were with the terrain in question. "Rocks move," he would say. "Mountains change, even in just a few years."[5] Rock slides intrigued him because of these surveys around the site of the Frank Slide. It had occurred quite recently, 29 April 1903, between

Bridgland's first and second summers in the mountains. Seventy-four million tonnes of rock had crashed down Turtle Mountain, burying a mine shaft and the eastern outskirts of the town of Frank, sweeping 1.5 km (about a mile) from the base of Turtle Mtn across the valley of the West Oldman (now Crowsnest) River, and claiming the lives of seventy people. Aside from the survey data he collected, Bridgland took mental notes of the geological characteristics of Turtle Mountain and made a point of avoiding climbs on mountain slopes he suspected might succumb to the same fate.

For any purpose they saw fit, members of the ACC could consult a voluminous hand-written index to the club's collection of photographs, organized according to subject and the names of mountains. The Whyte Museum of the Canadian Rockies in Banff holds a copy of an index of over thirty-two hundred photos taken and collected by Bridgland personally and added to the ACC's holdings. As the Photographic Committee foresaw in 1918, new uses for the photographs would emerge as time passed, but it could not guess what all of these would be: "[t]he Committee deem it unwise to place any restriction on the contribution of prints. Any and all will be welcome, even when they may seem to the taker to be only of passing interest. Even camp scenes and groups of climbers may acquire a value in the future of which we can have no conception now."[6] (Because of the loss of so many young men in the First World War, personal photographs resonated particularly deeply at the time.) Nearly a century later, the foresight of the ACC photographic archivists and the efforts of mountain photographers and surveyors generally have proven to hold potentially immense historical, cultural, and ecological value. The emphases of the Photographic Committee remind today's reader of Canadian culture's enduring anxiety over the sheer size of the country, and, as a response to that anxiety, its steady resolve to make inventories of it in one form or another.[7]

NEARLY A CENTURY
LATER, THE
FORESIGHT OF THE
ACC PHOTOGRAPHIC
ARCHIVISTS AND THE
EFFORTS OF MOUNTAIN
PHOTOGRAPHERS
AND SURVEYORS
GENERALLY HAVE
PROVEN TO HOLD
POTENTIALLY IMMENSE
HISTORICAL,
CULTURAL, AND
ECOLOGICAL VALUE.

✳ *The Rocky Mountain Repeat Photography Project* ✳

IN 1996, eighty-one years after Bridgland completed his maps of the Jasper survey of 1915, Jeanine Rhemtulla, a graduate student in the Department of Renewable Resources at the University of Alberta, signed up as a volunteer with Eric Higgs, then a professor in the Department of Anthropology, to work with the Culture, Ecology, and

Restoration (CER) Project.[8] This was an interdisciplinary initiative that aimed to study the diverse facets of ecological restoration in Jasper National Park (JNP). Higgs challenged his team of interdisciplinary researchers to come up with information to describe what the park looked like around the time when it was established as a forest reserve in 1907. Even to think about restoring an ecology to a former state, one must have accurate historical knowledge about ecology and about human interaction with it. The landscape, after all, is the result of both human and natural forces acting in combination.[9]

For her part, Rhemtulla decided to study change over time in the patterns of vegetation in the CER's sixteen-km-long study area of the upper Athabasca River valley.[10] So she set out in search of information about the state of vegetation in the valley in early times against which she could compare the data she would collect that summer. At the park's Interpretive and Tourist Information Centre, housed in the old stone superintendent's residence and offices that the government built on Connaught Drive in Jasper the year before Bridgland's survey, Cultural Resources Officer and Warden Rod Wallace pointed her towards a collection of 735 black-and-white photographic prints from 1915 that was stored in the basement of the centre.[11] Rhemtulla soon confirmed Wallace's thought that these photographs would be of great interest to her colleagues on the project: not only were they systematically taken and comprehensive in coverage but also they were unparalleled by any other early historical records of the area.

She brought them back to the CER group and, with its members, pondered and discussed how to establish a comparative study under contemporary circumstances and conditions. She and Higgs decided to try rephotographing a selection of the 1915 photographs from their original camera stations. After locating an aging large-format Linhof camera from the university's stores and learning how to use it, they set out with photocopies of Bridgland's pictures from one of his few low-level camera stations of the 1915 season, the one he had called Snaring. They hoped to locate the spot from which the photos were shot. With a little guesswork and considerable luck they soon found the station within a few metres of where they parked their vehicle by the railway bridge over Snaring River, a tributary of the Athabasca downstream from the town site of Jasper. (Fleming's railway surveyors called it Snare rather than Snaring.) On initial inspection, the photographs they

EVEN TO THINK ABOUT RESTORING AN ECOLOGY TO A FORMER STATE, ONE MUST HAVE ACCURATE HISTORICAL KNOWLEDGE ABOUT ECOLOGY AND ABOUT HUMAN INTERACTION WITH IT.

Photograph by M.P. Bridgland of Snaring River at the railway bridge, Stn 60, no 482, direction north, 1915.
Courtesy Jasper National Park.

took there revealed a remarkable difference in vegetation cover between 1915 and 1996.

By the end of the 1996 summer field season, Rhemtulla and Higgs had concluded that a larger-scale rephotography project would make a viable and valuable contribution both to the goals of the CER Project and to her thesis. Over the winter, they investigated uses and methods of repeat photography; then in the summer of 1997, returning to Bridgland's survey stations overlooking the project's chosen study area, they rephotographed the same views. Neither had had any previous interest in mountaineering, the history of photography, or surveying, so the learning curve for the two researchers was as steep as the mountainsides that awaited them. Yet as they began to climb and repeat the photographs, they were both badly bitten by the alpinism bug, deciding to re-shoot all 735 photographs at the ninety-two stations that Bridgland and his teams had occupied in 1915. They dedicated the summers of 1998 and

Photograph by J.M. Rhemtulla and E.S. Higgs of Snaring River at the railway bridge, Stn 60, no 482, direction north, 1999. Courtesy Rocky Mountain Repeat Photography Project, University of Alberta, University of Victoria.

1999 to this work. Thus was born what is known today as the Rocky Mountain Repeat Photography Project, an initiative that now occupies an interdisciplinary group of professors, librarians, graduate students, park staff, archivists, and other research collaborators.[12]

The practically-minded Bridgland would doubtless have been impressed by the repeat photography project's economy, for Rhemtulla and Higgs plus one assistant accomplished in two summers nearly all the climbing that his and Hyatt's/Norfolk's teams had done in one. Moreover, they did more bushwhacking to get to the foot of mountains since JNP has considerably more vegetation than it had in 1915, when fires were not prevented or suppressed. Infrequent lifts from benevolent helicopter pilots occasionally offered respite from the gruelling work, and a vehicle and roadways could whisk the rephotographers along the main valley closer to the foot of many mountains than the train could manage for Bridgland's party, but the meagre budgets of the 1990s for academic research precluded more than one packer-assistant, any beasts of burden, or a cook.

As Higgs and Rhemtulla looked further into the history of Bridgland's photographs, they found the accompanying maps that indicated the location of the camera stations, as well as Bridgland's booklet, *Photographic Surveying.* At about half the stations, part or all of Bridgland's huge rock cairns remained to mark the central camera station, but where these were absent the map and the booklet helped them both understand Bridgland's methodology and guess where exactly the camera had been set up. Alas, the transit notes that Bridgland recorded so diligently in the field, which detailed all the locations of the camera and transit stations, could not be found, but not for lack of trying: on at least three occasions researchers went to Ottawa to scour the National Archives (now Library and Archives Canada) and other federal repositories for any sign of them or any other pertinent data. But the elusive three-by-five-inch hardcover books remained hidden until 2002, well after the information they contained could have benefited the fieldwork of rephotography.

The lack of precise data locating the stations amounted to one of a number of Higgs and Rhemtulla's worries. They encountered and overcame innumerable obstacles during the one preliminary and two "full-bore" summers spent in the field. Aside from the usual perils and challenges associated with mountaineering, they met administrative and funding challenges, a shortage of time, and the difficulty of finding an engaged and devoted assistant. In 1998, one careless assistant ruined several days' worth of work by leaving a pack of exposed film in the rain. The assistant quit in mid-season and the summer's work looked as though it would fall far short of its goal. However, Bridgland's loss of Hyatt when he drowned in Lac Beauvert and the time taken from the survey by M.P.'s return journey to Revelstoke for the funeral kept the rephotographers' disappointment in perspective. Blessings always seem to come when most needed: two Californians arrived in Jasper and asked Higgs for volunteer work. Agreeing to room and board in return for their assistance, they proved to be diligent workers. Apparently born in mountaineering boots, experienced, and safety-conscious, they rendered the remaining weeks a success.

So the summer of 1999 began on schedule and only the occasional setback delayed the work. An assistant who was asked to repair a map unfortunately chose double-sided tape for the job. Higgs discovered the

AS HIGGS AND RHEMTULLA LOOKED FURTHER INTO THE HISTORY OF BRIDGLAND'S PHOTO-GRAPHS, THEY FOUND THE ACCOMPANYING MAPS THAT INDICATED THE LOCATION OF THE CAMERA STATIONS, AS WELL AS BRIDGLAND'S BOOKLET, *PHOTOGRAPHIC SURVEYING.*

error in the helicopter en route to a camera station. The rental of helicopters that summer cost an average of $800 per hour, so an anxious and distracted Higgs found himself trying to direct the pilot to their destination as quickly and efficiently as possible while wrestling with an unwieldy and torn map. Financially, the work remained precariously circumstanced all project long. Enough people saw value in it, however, and came to the rescue at critical points. A few guardian angels on the park's staff kept an eye out for free lifts in one of several helicopters being used to ferry materiel for the backcountry trail crews or standing by on fire patrol. Weldwood of Canada Ltd., based in Hinton, provided several hours of helicopter time. The value of all this was enormous: even a few hours in a chopper can save several days of footing it.

Higgs and Rhemtulla learned quickly that all helicopter pilots are not created equal. The work they asked the pilots to do, which involved touching down on narrow ridges buffeted by uncertain winds, created challenges that only an experienced and very calm pilot could accomplish. They spent a couple of frustrating hours in the air with inexperienced pilots burning up precious flight time, unable to touch down on the tantalizingly close peaks. Great good fortune came in the form of John Bell, a senior and revered mountain pilot. He took a special interest in the repeat photography project and bent whatever rules he could to contribute to it. On one memorable flight in late August 1999, Higgs remembers John's remarking of an especially stingy landing point atop Mt Aeolus, "You want me to land there? There is no there there!"[13] On the last day of the entire project's fieldwork, he estimated that he'd had to make three of the ten most difficult landings of his entire twenty-five-year career.

On that last day, the rephotographers had to locate not one but two stations used by Bridgland. In the end they had to visit the first on Friday, 27 August, and the other on Sunday. In her diary's entry for the second, Rhemtulla expresses herself poignantly in several respects: "John the wonder pilot dropped us on a precarious, narrow and unforgiving ridge. If my estimation of Bridgland's mountaineering was high before, it has reached excelsior heights today. This is a formidable mountain. The weather is good, but for how long this is difficult to estimate."[14]

Aside from the comedy of errors that seemed to unfold with each new batch of hired help, logistics remained as great a concern as

GREAT GOOD FORTUNE
CAME IN THE FORM
OF JOHN BELL, A
SENIOR AND REVERED
MOUNTAIN PILOT.

Photographs of Higgs and Rhemtulla occupying two of the ninety-two stations of M.P. Bridgland's survey of the central portion of Jasper National Park in 1915. Photograph one shows Eric Higgs setting up a mountaintop repeat photograph on Mt Henry, Fiddle Peak, Stn 53. Photograph two captures Jeanine Rhemtulla at work atop Roche Bonhomme, Stn 42, with other peaks of the Colin Range in the background. Photograph three shows the camera and tripod ready for work on Roche à Bosche, Stn 68. Courtesy Rocky Mountain Repeat Photography Project, University of Alberta, University of Victoria.

finances. Unlike in 1915, in the 1990s much had to be coordinated before and while research was done in a national park. After securing research permits and wilderness passes, the team had to assemble the right equipment at the right time as well as try to engage the interest of park staff immersed in their own work, reeling from deep cuts inflicted on their research budgets, and so for whom the project could not be accorded priority. It was often difficult or impossible to arrange logistical support such as conveniently-sited wilderness base camps. Then, like Bridgland, they had to hope for at least fair if not good weather. Snow plagued the team in 1998, while in 1999 long periods of rain and bad weather were followed by scorching hot days that lasted interminably. Despite the unrelenting heat, the shortened season required them to work every hour they had the energy to do so. Equipment also posed problems, particularly regarding the camera, which was found to have light leaks that Rhemtulla resourcefully detected with a flashlight and repaired with something Bridgland would doubtless have envied: duct tape. Meanwhile, the vagaries of the terrain remained a constant. On the climb up Mt Maccarib, Higgs wrote in his journal that "the peak is decomposing as I write. This changes my previous view about the short term stability of these mountains." This remark echoes Bridgland's warning to his assistants that no rocks in the mountains should be considered permanent fixtures.

Unlike Bridgland's teams, Rhemtulla and Higgs typically followed rather than blazed their own trails. "It was like having a conversation with Bridgland," she remarked during an interview in the summer of 2002. On the mountain peaks she found herself asking, "Where were you? What were you thinking?" Higgs agrees that they felt Bridgland's presence often: "We knew that we were standing in exactly the same place where he'd been eighty years before, and sometimes we were only the second person to ever stand there. It was both awesome and eerie." Through such affinities they began to empathize with him even though they then knew about him only what his maps, photographs, and official publications disclosed. As well, they knew he must have been one superb mountaineer and, moreover, must have had exceptional leadership skills to inspire his men, year after year, to keep up their momentum as good conditions and weather waxed and waned. The qualities of discipline and organization feature strongly in the various records; so although the exact positions of stations were sometimes elusive, in the end, Bridgland's choices invariably made sense to the rephotographers. In the few instances where the location of a camera station struck them as bizarre, they concluded that weather had dictated his choice, as indeed it sometimes would theirs.

As they worked to find each location, they knew that M.P. would have taken far less time to compose his views, despite his challenges in levelling and centring the camera. Whereas he could frame the land-

"WE KNEW THAT WE WERE STANDING IN EXACTLY THE SAME PLACE WHERE HE'D BEEN EIGHTY YEARS BEFORE, AND SOMETIMES WE WERE ONLY THE SECOND PERSON TO EVER STAND THERE. IT WAS BOTH AWESOME AND EERIE."

scape according to a surveyor's aesthetic, the rephotographers aimed to recapture the choices made by their predecessor. So they had to line up more than just the prominent features of the landscape; they had to make the same photograph that Bridgland had made. Theirs was a dual focus, first on repeating Bridgland's photographs exactly, then on observing and analyzing the changes in the landscape over time. Painstakingly, they sometimes took hours aligning fine-grained foreground features—rocks, vegetation, cairns—with the versions of them on the mountaintops today. Most difficult were the survey stations that had undergone major changes through human activity or rock fall. The same location eight decades later was sometimes difficult to find and fascinatingly altered. However, the photo from atop The Whistlers is intriguing for the continuity of the foreground, for although most of the rocks lack the vegetative growth that they possessed in 1915, probably owing to amount of human traffic near the top of the gondola that takes tourists up the mountain, a great many of the ones remain in place and in sufficiently similar relation to each other as to afford ready recognition in the pair of photos. That said, many of the rocks easier for a person to lift have been disturbed. At Surprise Point, the lack of a large slab of rock is noticeable but the face of the left foreground again looks recognizable. But the rephotograph masks the effort involved in finding/recognizing the view from Bridgland's station. The viewer of the pair of photos is, of course, looking at the greatest resemblance in the landscape between 1915 and 1999, but the rephotographers' efforts to find that resemblance, or what is left of it, is lost except in their memories of the day's work. Where the rephotograph goes silent, Higgs rehearses the experience of surmounting this relatively low peak at the south end of Tonquin Valley in order to make it: it "was the stuff of fireside stories—backtracking on our original route to ascend a steep, hard-packed couloir, finding the rock on the summit section treacherous and mobile, and then spending an inordinate amount of time determining the specific location of the photos before realizing that several key rocks, and large ones, had disappeared over eighty years. The descent was harrowing, too, and we ended up on the other side of the mountain having to circumnavigate the base to find our camp." Higgs and Rhemtulla demurred when invited to elaborate on the company that mosquitoes provided throughout their sojourn—perhaps the most disagreeable surprise of Surprise Point.[15]

THEIRS WAS A
DUAL FOCUS, FIRST
ON REPEATING
BRIDGLAND'S
PHOTOGRAPHS
EXACTLY, THEN
ON OBSERVING AND
ANALYZING THE
CHANGES IN THE
LANDSCAPE OVER
TIME.

Photograph by M.P. Bridgland of Mt Fraser from Surprise Point, Stn 14, no 115, direction southwest, 1915.
Courtesy Library and Archives Canada R214.

Photograph by M.P. Bridgland of
Mt Fraser from Surprise Point, Stn
14, no 115, direction southwest, 1915
(detail). Courtesy Library and
Archives Canada R214.

Photograph by E.S. Higgs and J.M. Rhemtulla of Mt Fraser from Surprise Point, Stn 14, no 115, direction south-west, 1999. Courtesy Rocky Mountain Repeat Photography Project, University of Alberta, University of Victoria.

Photograph by E.S. Higgs and J.M. Rhemtulla of Mt Fraser from Surprise Point, Stn 14, no 115, direction southwest, 1999 (detail). This considerable change goes some way to attesting to the name, "choss heaps," that those mountains in the Rockies with a preponderance of limestone sometimes attract from alpinists who find a mountain seemingly falling apart as they climb. Courtesy Rocky Mountain Repeat Photography Project, University of Alberta, University of Victoria.

Certain modern provisions distinguished the repeat photography from the surveyor's. Most importantly, the rephotographers had the advantage of maps. Every day presented new terrain to Bridgland, often terrain that had never been documented in any preserved manner. By contrast, Rhemtulla and Higgs could study pictures each night to rehearse what views they would have to try to find and then duplicate the next day. The fact that they enjoyed modern amenities rendered the mountain work not so much easier as very different, and yet the goal of 735 rephotographs aligned them intimately with Bridgland's project. Certainly their boots were less heavy, more waterproof, and more tailored, but the mountains still had to be ascended with care and concentration, and the same equipment that men hauled up them for the photography had to be taken up under much less manpower for the rephototopography. While they gained tremendous respect for Bridgland's accomplishments, skill, and fortitude, they recognized as well that he benefited from a larger support team, relatively less dense and extensive forests through which to travel to the foot of mountains, and, by 1915, thirteen years' experience in mountaineering and photographic surveying.

The trip up Roche Ronde on 22 August 1999 illustrates the arduous nature of the repeat survey for Rhemtulla and Higgs. After a 5:30 a.m. start from the Palisades Centre, their summer headquarters east of Jasper town site, the team drove for a half-hour along the winding Celestine Lake Road to an access point close to the CNR line, and then walked along the tracks for thirty minutes before fording a chest-high flooded impoundment to begin the ascent. The first 3,500 feet involved scrambling through dense bush, along serpentine ridges, and eventually to a scree gully and cliff that separated the peak from the climbers by fifty feet of mid-grade technical climbing (roughly 5+ on the scale approved by the UIAA [Union Internationale des Associations d'Alpinisme]). Thus, a short pitch of relatively unexciting technical climbing requiring a sizeable kit of mountaineering equipment took six hours of nasty whacking and scrambling.

Higgs and Rhemtulla's ascent of Mt Utopia (8,537 ft [2,602 m]) offers one of the better illustrations of how widely the rephotographers' experience on any single mountain could differ from the Dominion Land Surveyor's. It will be remembered that because in August 1915 M.P. encountered a luxuriously verdant meadow while surveying Mt Utopia, he named the peak accordingly. His successors

THE FACT THAT THEY ENJOYED MODERN AMENITIES RENDERED THE MOUNTAIN WORK NOT SO MUCH EASIER AS VERY DIFFERENT, AND YET THE GOAL OF 735 REPHOTOGRAPHS ALIGNED THEM INTIMATELY WITH BRIDGLAND'S PROJECT.

Photograph by M.P. Bridgland of Mt Utopia from Sulphur Skyline (above Miette Hot Springs and Sulphur Creek), Stn 84, no 686, direction southwest, 1915. Courtesy Library and Archives Canada R214.

Photograph by J.M. Rhemtulla and E.S. Higgs of Mt Utopia from Sulphur Skyline (above Miette Hot Springs and Sulphur Creek), Stn 84, no 686, direction southwest, 1999. Courtesy Rocky Mountain Repeat Photography Project, University of Alberta, University of Victoria.

set out early on a gorgeous day but soon found themselves climbing in a cloud bank that persisted throughout their time on the mountain. What had looked from the map to be a relatively straightforward day turned gradually into weather-wracked torment. Weather is a neutral word to the man in the street; to alpinists it almost invariably carries negative connotations. Clouds thickened and the thermometer plummeted. They reached a plateau near the summit and stopped for lunch. It started to snow. They vainly hoped it would clear; the forecast had promised fine weather, after all. After lunch they had only 500 ft left to tackle, about twenty minutes' worth of climbing. Up they climbed, carefully but uneventfully, from the plateau to the cloud-filled summit, where any effort at photography proved pointless. On their return to the plateau, however, they ended up more than two hundred feet below the critical junction. The safest remedy required re-ascent in order more carefully to retrace the route to find the original path, or at least to locate an alternate route that steered away from treacherous cliff bands. There was enough repeating going on in the project already without this! By then, however, the visibility had diminished to a mere twenty-five feet. Higgs remembers this as their single worst experience on a mountain. Had Utopia conspired against them? It seemed so, for when they reached bottom the sun shone. A large cloud hovered over the trickster as they drove along the highway back to Jasper but it was the only one in the sky! Surely utopias are not so wily and unwelcoming. Later in the summer, they were able to capture the views they needed but only by being deposited on the peak by John Bell's helicopter.

✳ *Comparing the Original and Repeat Photographs of Jasper* ✳

CHANGE IN THE LANDSCAPE was easier to recognize than constancy—and it was more dramatic. Difference is what jumps out at anyone who views pairs of original and repeat photographs of Jasper. As Higgs writes, "Constantly changing ecosystems influenced by layer upon layer of human activity had created a landscape of difference. Our work froze two moments in time, and from this we can begin the laborious process of interpolating and extrapolating meaning: What did the landscape resemble in the nineteenth century and before? What are the clues and signals from 1915? What do the changes from 1915 to the present chart for the future?"[16] Change creates a story so powerfully compelling that it displaces any other

DIFFERENCE IS WHAT
JUMPS OUT AT ANYONE
WHO VIEWS PAIRS OF
ORIGINAL AND REPEAT
PHOTOGRAPHS OF
JASPER.

interpretation of the landscape. The repeat photographs concern the study of change, or else why would they be taken? Contemporary society is infused with evolutionary perspectives and most people have incorporated the belief that people and nature are constantly changing, and on a particular trajectory. This is the story underlying geologic formation, plant succession, and cultural beliefs, to name a few examples. At the same time, wilderness areas such as Jasper National Park are assumed to be primeval regions beyond the grip of human-induced change. Because wilderness seems to offer the antidote to technological excess, national parks have come to be regarded as so very precious to humanity.

Paradoxically, we exempt wilderness from the demands we make of the rest of the non-human world, believing that the regions on which we confer wilderness status can be shielded from the flow of time and eruptions that evolution and social activity bring about. In most respects this fixed notion of wilderness is simply untrue. Showing that it is, the Bridgland and repeat photographs taken together have much incontrovertible testimony to offer. It is easy to swing too far in either direction, that is, either towards the view that landscapes are in constant change and thereby undeserving of restorative activities (this line of reasoning asks, Why restore damage if we know the landscape will change in any case?), or towards the view that a place such as JNP is an ageless and unchanging monument. A museum even. Meanwhile, though, it is also possible that the need to dispel the myth of wilderness as untouched and unchanging prompts in us too much emphasis on change. Might the photographs reveal as much about what has *not* changed in the interval as what has?

A good example might be the pictures made from a low-lying station, the one at Snaring River bridge, mentioned earlier as revealing a remarkable amount of change over more than eight decades. In order to amass a set of photos that was restricted to a short span (just the sixteen-month span covered by the two summers of 1998 and 1999), Rhemtulla decided that she had to re-shoot the photos that she had taken first in 1996. So on 6 July 1999, "a wonderful sunny day," she drove down the road to the Snaring overflow campground. Her notes from that day tell us not only something more of the work involved in rephotography but also something of both the endurance of some features in the landscape and the changes in others, both human and non-human, since Bridgland's day:

PARADOXICALLY, WE
EXEMPT WILDERNESS
FROM THE DEMANDS
WE MAKE OF THE REST
OF THE NON-HUMAN
WORLD, BELIEVING
THAT THE REGIONS
ON WHICH WE CONFER
WILDERNESS STATUS
CAN BE SHIELDED
FROM THE FLOW OF
TIME AND ERUPTIONS
THAT EVOLUTION AND
SOCIAL ACTIVITY
BRING ABOUT.

Location 1

It never ceases to amaze how clearly you can still see the remnants of the retaining walls along the Snaring. All overgrown, of course, but with a little bit of poking, you can still see bits of the great big timbers along the rock piles. The next two photos were taken some place just SW of the south end of the Snaring Bridge. It's hard to tell exactly where because of all the silverberry, willow and young balsam poplar. Along the river edge. I'm going to take the first 2 photos in a spot close to the original, and then move forward to the water's edge so I can get some pictures where [beyond the vegetation that obscures the view from Bridgland's station] I can actually see something. About 5 metres south and 1 m west of the SW end of the car bridge over the Snaring. The "end" of the bridge in this case is the edge of the green metal structure on the SW side.

Location 2

Moved down to the water's edge so that the photos will show something other than Eleagnus. Moving 12 metres forward, parallel to bridge, approximately north still about 5–6 metres west of the bridge. See Rollei photo #2–1 taken so that the tripod and I form a perpendicular line to the bridge.

Location 3

On the extreme SW edge of the bridge, camera as close to the steel edge as possible. Actually, I bet Bridgland took both of the next 2 photos from the same spot on the bridge edge, but because the bridge has changed and now has a high post structure which obliterates the view, I'll have to be a bit arbitrary in my camera locations.

Location 4

I'm on the other side (east) of the bridge now. I can't take the photo off the edge of the bridge, trees block the view. So, I've moved to the first large hole in the bridge lattice structure about 8 metres from the SE end, right up against the metal railing.[7]

The bridge and, as one can clearly see in the photo (on page 217), the trees on the far bank and on the mountain in the background off the

north end of the bridge have changed enormously, but signs endure of
the retaining wall built by the railway workers, the skyline is readily
recognizable of course, and the river still takes a similar course, differences in water elevations notwithstanding. The vast change in the size
and type of trees and other flora suggests that it is a changing landscape because it lies inside a park, but for the same reason it is a consistent landscape: no hydro dam has changed the river's course, no
industrial development has altered the quality of its water; the bridge
has been rebuilt but no change in its location has been permitted and
no railway debris is allowed to accumulate. This is what the naked eye
can see; inspect the digitalized version of the pairs of photos and
zoom in on particular details, and one can see a great deal more. Here
is an advantage over the technology available in Bridgland's day.[18]

For Rhemtulla, the photography marked just the beginning of her
thesis project. Digitalization of the photos was not a priority in 1999.
The next step was to find a way to use the pairs of images. Qualitative
analyses of the repeat photographs were well and good—any one could
compare a photograph from 1915 with one taken eight decades later—
but she could not straightforwardly compare the vegetation cover in
two photographs until she determined that photos shot on an oblique
angle from mountaintops (that is, on an angle down and across a landscape rather than, as in aerial photography, from directly overhead)
showed more vegetation cover than truly existed. How distinctive, she
wondered, were the vegetative compositions shown in each source she
could access: Bridgland's record from 1915, a series of aerial photographs taken in 1949, and her and Higgs's rephotographs? In short, she
needed to determine if photos shot on an angle rather than from
directly overhead could serve reliably as the basis for quantitative
analysis.

After reading up on Bridgland's technique for composing his maps
from the photographs, she realized that she would need more time to
complete her analysis if she was to match his standard. Determining
how to automate the process or to create software to do so lay beyond
the scope of her thesis; instead, she decided to draw polygons around
distinctive vegetation types on the photographs and suppose their
perspective was aerial rather than oblique. Before adopting this method,
she determined that placing polygons over a set of aerial photographs
taken in 1991 of the study area and comparing them to polygons placed
over the oblique rephotos of the same area that she made on a trial

THE VAST CHANGE IN
THE SIZE AND TYPE
OF TREES AND OTHER
FLORA SUGGESTS THAT
IT IS A CHANGING
LANDSCAPE BECAUSE
IT LIES INSIDE A
PARK, BUT FOR THE
SAME REASON IT IS
A CONSISTENT
LANDSCAPE.

basis in 1997 revealed that a reliable basis of comparison was readily possible with respect to quantitative analysis. That is, the vegetation cover appeared if not identical then sufficiently similar in aerial and mountaintop photographs from the same decade as to warrant quantitative analysis of vegetation in oblique photographs from different decades. This method worked satisfactorily in concert with topographic maps of the study area for she could estimate how the contours of her polygons translated onto the map. In the spring of 1999, she reached the conclusion that the vegetation had changed significantly over eighty years: forest cover had grown increasingly homogenous, more widely distributed, and more dense.[19]

THE SYSTEMATIC,
COMPREHENSIVE
NATURE OF
BRIDGLAND'S WORK
OPENED THE
POSSIBILITY OF
STUDYING VIRTUALLY
ANY DIMENSION OF
THE LANDSCAPE
VISIBLE IN THE
IMAGES.

Rhemtulla's research pointed a way forward for the project. By deriving explicit quantitative measures from the photographs, she showed the Bridgland and repeat collections to be more valuable research and management tools than most conventional photographic collections. The systematic, comprehensive nature of Bridgland's work opened the possibility of studying virtually any dimension of the landscape visible in the images: human activities, movement of water, vegetation, geological structures, snow pack and glaciers, and even miniature details in the foreground of the photos such as changes in the growth of lichens on rocks. The photos are a gold mine for extracting insights about the ways in which a precious landscape such as Jasper's main valleys has changed over time in response to intensifying human activity, shifting climate, and inexorable ecological processes such as vegetation succession, wildfires, windstorms, insect outbreaks, and floods. To exploit this promise further, work is under way to develop software that will accomplish what Rhemtulla originally wanted: a reliable, accurate, and effective way of quantitatively comparing various landscape features.

✳ *Battling the Scythe of Old Man Time—*

Preserving Both Sets of Photographs ✳

OVER THE COURSE OF THE REPEAT PHOTOGRAPHY PROJECT, Bridgland's pictures continued to grow on Higgs and Rhemtulla. They developed increasing respect for the difficulty involved in making them. Moreover, not yet knowing that the glass plate negatives from which they were made still existed, they esteemed the prints all the more highly. Of course, they knew of and had worked from the collection held by JNP and, after a reassessment, transferred from the Interpretive Centre to the JNP Library (also located in Jasper), but they had since discovered that the Jasper-Yellowhead Museum and Archives had copies of the same four-by-six-inch contact prints made from the glass plate negatives, bound in groups of roughly fifty photographs to a book. Had other sets of the photographs survived? And what *had* happened to the original glass-plate negatives? Several searches through national and provincial archives suggested only that they were presumed to have been destroyed. Edgar Bridgland had been contacted; he knew of their whereabouts up until about 1992 but not thereafter. If the original negatives had, like the transit notes, disappeared, surely measures had to be taken to ensure that the same fate did not befall the rephotographic work. It was decided that the Bridgland pictures, paired with their repeats, should constitute a digitalized collection of images preserved on a website from which studies of changes in landscape and vegetation could be undertaken.

Jenaya Webb, a master's student in the Department of Anthropology at the University of Alberta, spent the summer of 2000 creating a digital record of both sets of photographs.[20] She scanned the pictures at a high resolution, one which would allow the digital images to be viewed without distortion in magnification. For example, in one view of the Overlander Trail, a Métis family's farm can be seen. It had been abandoned in 1910, five years before Bridgland did his work, when all Métis homesteaders were removed from the upper Athabasca River valley. The farm buildings are discernible only in the digital magnification of the four-by-six-inch print, not in the print itself; what the computer monitor can reveal in the photograph, the photograph itself "hides."

THE FARM BUILDINGS ARE DISCERNIBLE ONLY IN THE DIGITAL MAGNIFICATION OF THE FOUR-BY-SIX-INCH PRINT, NOT IN THE PRINT ITSELF.

Aside from the visual changes that a careful comparison of the two sets yields, Bridgland's notes imply that he oriented his camera in a general sense only, that is, approximately in the stated direction, north-northeast, for example, and depended for desired levels of accuracy on his triangulation measurements with the much more exact transit. So, working from his stated directions on the photographs can be somewhat misleading at times, and the disorientation can be compounded by the increase that has occurred since his day in the declination of the compass as a result of the gradual and recently quite rapid shift in the position of Magnetic North. So the two sets of photographs attest to the evolution of the terrain over time with respect to systems of orien-

Photograph by M.P. Bridgland of Athabasca River valley, mouth of Snaring River, and Colin Range from The Palisade, Stn 59, no 471, direction eastnortheast, 1915. Courtesy Jasper National Park.

tation that we tend to think of as standard. They are portraits of the landscape at a moment in time, not the permanent record that Bridgland and the DLS as a whole tended to regard them as being. The researchers with the Rocky Mountain Repeat Photography Project have to keep reminding themselves that their sets of photos are not permanent either.

In a perfect world, all pertinent records would have come to light before the rephotography project began in JNP. In the event, not until 2002, well after the creation of the digital repository, did researchers' efforts at last bring them to light. This was the work of Gabrielle Zezulka-Mailloux and Trish Bailey, graduate students in University of Alberta's English and Anthropology departments, respectively. Boxes of professional correspondence arranged in indexed files in chronological order helped piece together the history of Bridgland's career in the mountains; transit notebooks found in a mostly unused vault at the

Twenty-five per cent enlargement of portion of no 471,
showing the Athabasca River.

Geodetic Survey in Ottawa confirmed the accuracy of Rhemtulla and Higgs' location of the original 1915 survey stations; the glass plate negatives found uncatalogued, reposing in cardboard boxes in a warehouse belonging to the National Archives rewarded undying if gradually flagging hopes that they had somehow eluded the dustbin. Each of these records contributes to our understanding and appreciation of the scope and achievement of Bridgland's work. Their emergence coincided with the surfacing of some details of family history, collaborators, and friendships.

Serendipity generated the first views of the photos by Rhemtulla and Higgs in 1996, but it was their interest in systematic technique and scientific insight that prompted them to conduct the repeat photography project and other University of Alberta staff, such as Sandy Campbell, David Cruden, and Ian MacLaren, to collaborate with them.

One hundred per cent enlargement of the same portion, revealing the buildings of homesteaders Isidore Finley (*upper [east] bank*) and Adolphus Moberly (*lower [west] bank*), who had been compensated and removed from the valley in 1910, prior to its becoming Jasper Park in 1911.

In doing so they aligned themselves with an American scientific tradition stretching back to the 1960s of using photographs to study landscape change.[21] The Jasper project was one of the first major Canadian experiments in repeat photography and likely one of the largest undertaken anywhere.

The surprises that emerge are an enjoyable aspect of research. None of the original researchers would have guessed, for instance, that Bridgland and his teams' work in the summer of 1915 would turn out to be the tip of an enormous iceberg of photography. This discovery led to a formal partnership between Library and Archives Canada and the researchers—led now by Higgs at the University of Victoria—to conserve and make accessible the entire collection of images, which may prove to be the largest systematic national collection of oblique photographs. Along the way, researchers from many disciplines have joined the RMRPP, including specialists in geomatics, paleoclimate, vegetation ecology, history, anthropology, photography, and resource management. The office of the Surveyor General in the person of Alec MacLeod has proved to be a tireless source of assistance and information. Several graduate students have completed projects based on the photographs and more will finish in coming years. Higgs is now joined by Trudi Smith, a professional photographer and graduate student, graduate student Adrienne Shaw, and Parks Canada Warden Rob Watt, in Waterton Lakes National Park, to complete repeat photography of Bridgland's 1913 and 1914 phototopographic survey of that region. Jeanine Rhemtulla, now a research associate of the RMRPP, continues her interest in historical ecology and is pursuing doctoral studies at the University of Wisconsin. The hope is that the repeat studies will present another systematic layer of understanding landscapes and how they change. Future researchers and managers will surely look back on the repeat surveys as a vital addition to the original surveys, so that the twentieth century may be looked upon more or less as book-ended by the work of Bridgland and his successors.

THE JASPER PROJECT
WAS ONE OF THE FIRST
MAJOR CANADIAN
EXPERIMENTS IN
REPEAT PHOTOGRAPHY
AND LIKELY ONE
OF THE LARGEST
UNDERTAKEN
ANYWHERE.

✳ *Bridgland's Legacy* ✳

ACCORDING TO MARGARET HESS, Bridgland would laugh about all this attention, but he would also be thrilled to know that people were taking an active interest in what he studied and loved. He, too, aimed to understand landscape change in the mountains, although as a

breaker of cartographical new ground he lacked the basis for comparison that the ongoing RMRPP enjoys. Instead, he looked at the rocks
in front of him and tried to figure them out. From the universal to the
particular, from astronomy and geology, to changes in glaciers and a
photographic inventory of alpine flowers, Bridgland's interests
extended from developing an understanding of the historical formation of the mountains to appreciating the flora and fauna alive amidst
them, the fossils that survived from long extinct species, and to a lesser
extent the peoples and settlements they accommodated. When he was
not busy mountaineering, surveying, photographing and developing, or
map-making, he was indulging his fascination with the botany, geology,
paleontology, and history of the Rocky Mountains, or was busy
imparting his knowledge to others in talks or publications. "He was a
very shy man who was not prone to talking about himself," recalls
Hess, "but if you were to ask him about one of the subjects he was
passionate about, he was hard to stop. His real hobby horses were the
retreat of the glaciers and wildflowers."[22]

From the earliest hand-drawn sketches that were turned into rough
topographic diagrams, to the photography that gave rise to the first
detailed topographic charts of the Rockies, map-making has always
involved looking carefully at terrain and rendering its three dimensions
into two on paper. Until late in the twentieth century, map-making
also involved looking at the skies overhead and describing the position
of landscape features relative to the stars' positions. Now we have satel-
lites looking from the stars' perspective down at the Earth. From the
images they retrieve, we can create highly accurate maps and photographs of any given place under the satellite's orbit. This sort of
mapping is called *remote sensing* precisely because the photographer or
ostensible surveyor has no contact with the land being mapped. Would
Bridgland have approved of the use of satellite imagery for mapping
purposes? As a truly professional surveyor, he probably would be in
awe of the accuracy attainable and would probably also delight in the
fact that individual contour lines no longer had to be plotted by hand,
using needles and thread or hair. Conversely, throughout his life he
engaged and encouraged others to engage in an active observation of
the world—to better see the land by touching it, living in it, challenging and being challenged by it, observing and focusing carefully.
He physically encountered nearly all the mountains for which he drew
the contours and he observed each of them not just through the lens of

THROUGHOUT HIS LIFE
HE ENGAGED AND
ENCOURAGED OTHERS
TO ENGAGE IN AN
ACTIVE OBSERVATION
OF THE WORLD—TO
BETTER SEE THE
LAND BY TOUCHING IT,
LIVING IN IT,
CHALLENGING AND
BEING CHALLENGED
BY IT, OBSERVING AND
FOCUSING CAREFULLY.

Photograph by T. Smith of a portion of the Rocky Mountain Repeat Photography Project's exhibit, *Vantage Points: Scientific Photography in Jasper National Park*, Jasper-Yellowhead Museum and Archives, 5 July to 26 October 2003. Courtesy the photographer.

Photograph by T. Smith of a portion of the Rocky Mountain Repeat Photography Project's exhibit, *The Rockies through the Lens of Time*, Library and Archives Canada, Ottawa, 15 April 2005 to 15 January 2006. Courtesy the photographer.

his camera but also through the awe and the subtlety of daily travel in, up, and down them.

Mountain surveying was difficult, taxing, repetitive, exacting, enthralling, delightful, rewarding work for Bridgland. This list sounds paradoxical, but most labours of love are just that. During the course of his quarter-century-long career he could marry his physical passion for mountaineering and his creative passion for photography, both of which nourished his intellectual desire to inquire into the natural world. His maps and articles on hiking trails and climbing routes leave to all of us tools with which to explore and learn for ourselves what it means to be engaged with one's environment. The public response to *Vantage Points* and *The Rockies through the Lens of Time*, two recent exhibits of his photographic work, demonstrate that ongoing engagement.[23] M.P. mapped the mountains in ways and to degrees that continue to engross, exhilarate, and enlighten us today. ✳

✳ Notes ✳

1 For example, during his quarter-century (1911–36) as the first commissioner of the Department of the Interior's Parks Branch, Harkin, Bridgland's near contemporary, was a strong, early voice advocating wilderness protection. See Mabel B. Williams, ed., *The History and Meaning of the National Parks of Canada: Extracts from the Papers of the late Jas. B. Harkin, first Commissioner of the National Parks of Canada* (Saskatoon: H.R. Lawson Publishing, 1957); and Gavin Henderson, "James Bernard Harkin: The Father of Canadian National Parks," *Borealis* vol. 5, no. 2 (Fall 1994): 28–33. Harkin's was not the lone conservationist voice but it took most of his career to turn parks from solely a profit-making initiative into a wilderness preservation enterprise:

> The national parks administration had grown out of a bureaucratic leviathan, the Department of the Interior, whose principal purpose was not the conservation of resources but their commercial exploitation. It took years, even decades, for a small group of dedicated conservationists within the Dominion Parks Branch to shrug off the commercialism that characterized their mother department and establish a unique set of principles that would guide national park development well into the 1960s, when they would be reinforced even more strongly. Those principles, which took the preservation of natural resource integrity as their point of departure, were embodied in the National Parks Act of 1930.

(Great Plains Research Consultants, "Jasper," 244.) See also Canada, Statutes, 20–21 George V, Chap. 33, "The National Parks Act," Assented to 30 May 1930; rptd. in *Documenting Canada: A History of Modern Canada in Documents*, ed. Dave De Brou and Bill Waiser (Saskatoon: Fifth House, 1992), 299–302; and C.J. Taylor, "Legislating Nature: The National Parks Act of 1930," *Canadian Issues/Thèmes canadiens, vol XIII, 1991: To See Ourselves/To Save Ourselves: Ecology and Culture in Canada*, ed. Rowland Lorimer, Michael M'Gonigle, Jean-Pierre Revéret, and Sally Ross (Montreal: Association for Canadian Studies, 1991), 125–37.

2   Cyril G. Wates, "Annual Report of the Photographic Committee," *Canadian Alpine Journal* 10 (1919): 114.

3   Wates, 115.

4   M.P. Bridgland, *Description of & Guide to Jasper Park*, 29.

5   Margaret Hess, interview with Gabrielle Zezulka-Mailloux, Calgary, Mar. 2002.

6   Wates, 116.

7   On the rise of the inventory sciences in Canada, see Suzanne Zeller, *Inventing Canada: Early Victorian Science and the Idea of a Transcontinental Nation* (Toronto: University of Toronto Press, 1987).

8   A precursor to the Rocky Mountain Repeat Photography Project, this project issued a final report: Eric Higgs *et al.*, *Culture, Ecology and Restoration in Jasper National Park: An Interdisciplinary Research Project* (Dept. of Anthropology, University of Alberta, 1999).

9   These insights are described and elaborated in Eric Higgs' recent book on the philosophy and practice of ecological restoration, *Nature By Design: People, Natural Process, and Ecological Restoration* (Cambridge, Mass.: MIT Press, 2003).

10 The 64 km^2 of the study area lies in the Athabasca valley on either side of the Yellowhead Hwy and Athabasca River, from about the large bend in the river and road after the mouth of Maligne River, in the south, north and downstream to a point slightly beyond the bridge across the Athabasca River, in the north, and including the valley walls. See Higgs, *et al.*, *Culture*, Fig. 4–1, 41.

11 It is likely that Bridgland himself was responsible for sending copies of his photographs to this office, to be kept or used by the administration as it saw fit.

12 The RMRPP's website is located at http://bridgland.sunsite.ualberta.ca/main/index.html

13 Measuring 8,672 ft (2,643 m), Mt Aeolus forms part of the Bosche Range that rises between the valleys of the lower Snake Indian River and Moosehorn Creek, near JNP's eastern boundary.

14 J. Rhemtulla, Diary for 1999 Field Season; courtesy the author; available at http://bridgland.sunsite.ualberta.ca/main/station.php?id=71&action=moreinfo
    John Bell died suddenly of a heart attack the following year, while dancing with his partner Joanne who, as an employee at Weldwood, took him to the

company's annual Christmas party. It was a tragedy to be sure, though perhaps not the worst way for a helicopter pilot to die.

15 Eric Higgs, Jeanine Rhemtulla, I.S. MacLaren, email correspondence, 2–3 June 2005.

16 Higgs, *Nature By Design*, 177.

17 Jeanine Rhemtulla, Diary for 1999 Field Season; courtesy the author; available electronically at http://bridgland.sunsite.ualberta.ca/main/station.php?id=60&action=moreinfo

18 The digitalized pair of photos may be found at http://bridgland.sunsite.ualberta.ca/main/station.php?id=60&action=

19 The published version of these findings is Jeanine Rhemtulla *et al.*, "Eighty Years of Change: Vegetation in the Montane Ecoregion of Jasper National Park, Alberta, Canada," *Canadian Journal of Forest Research* 32 (2002): 2010–21. See also Jeanine Rhemtulla, "Eighty Years of Change: The Montane Vegetation of Jasper National Park," M.Sc. Thesis, Department of Renewable Resources, University of Alberta, 1999.

20 These may be viewed online at the Rocky Mountain Repeat Photography Project website, at http://bridgland.sunsite.ualberta.ca/main/index.html which also comprises associated historical and scientific notes and articles about the photographs.

21 Repeat photography studies in North America have a long history but came to the fore in the 1960s with James R. Hastings and Raymond M. Turner, *The Changing Mile: An Ecological Study of Vegetation Change with Time in the Lower Mile of an Arid and Semiarid Region* (Tucson: University of Arizona Press, 1965). Subsequent studies include the following: Garry F. Rogers, Harold E. Malde, and Raymond M. Turner, *Bibliography of Repeat Photography for Evaluating Landscape Change* (Salt Lake City: University of Utah Press, 1984); William J. McGinnies, Homer L. Shantz, and William G. McGinnies, *Changes in Vegetation and Land Use in Eastern Colorado* (Washington, D.C.: U.S. Department of Agriculture [ARS-85], 1991); Thomas T. Veblen, and Diane C. Lorenz, *The Colorado Front Range. A Century of Ecological Change* (Salt Lake City: University of Utah Press. 1991); Mary Meagher and Douglas B. Houston, *Yellowstone and the Biology of Time: Photographs Across a Century* (Norman: University of Oklahoma Press, 1998); and Robert H. Webb, *Grand Canyon, a Century of Change. Rephotography of the 1889-1890 Stanton Expedition* (Tucson: University of Arizona Press, 1996). A recent overview of photographic and mapping techniques is available in Tina Reithmaier, "Maps and Photographs," in D. Egan and E.A. Howell, eds., *The Historical Ecology Handbook: A Restorationist's Guide to Reference Ecosystems* (Washington, DC: Island Press, 2001), 121–46. Application of this methodology to another continent is available in John Pickard, "Assessing Vegetation Change over a Century using Repeat Photography, *Australian Journal of Botany* 50 (2002): 409–14.

22 Margaret Hess, interview with Gabrielle Zezulka-Mailloux, Mar. 2002.

23 *Vantage Points: Scientific Photography in Jasper National Park; The Bridgland & Repeat Photography Projects*, curated Trudi Smith, Showcase Gallery, Jasper-Yellowhead Museum and Archives, Jasper, 5 July–26 Oct. 2003 (thereafter, archived on the Internet at http://bridgland.sunsite.ualberta.ca/display/); and *The Rockies through the Lens of Time: The Work of Morrison Parsons Bridgland and the Rocky Mountains Repeat Photography Project*, curated Trudi Smith, Library and Archives Canada, Ottawa, 26 Apr. 2005–31 Jan. 2006.

{ ROCKY MOUNTAIN REPEAT PHOTOGRAPHY PROJECT
UNDERSTANDING LANDSCAPES AND HOW THEY CHANGE

CHANGES TO THE GLACIERS AND MOUNTAINS
OF JASPER NATIONAL PARK

Photograph by M.P. Bridgland of Mt Edith Cavell and its "Ghost" glacier (today's Angel) from Cavell Meadows,
Stn 2, no 19, direction southsouthwest, 1915. Courtesy Jasper National Park.

✳

WHEN BRIDGLAND included an image of this mountain in his *Description of & Guide to Jasper Park*, he noted that "the shape and shadings of the glacier [to the left of center] of the Ghost [Angel] present the appearance of a flying figure with outstretched wings" (42). Only a much keener imagination could trace the same shape in the remnant of that glacier today.

Digitalization of Bridgland's photographs permits researchers to zoom in to particular features in order to study further particular alterations in the glacier or its composition in 1915:

Photograph by J.M. Rhemtulla and E.S. Higgs of Mt Edith Cavell and its "Ghost" (today's Angel) glacier from Cavell Meadows, Stn 2, no 19, direction southsouthwest, 1999. Courtesy Rocky Mountain Repeat Photography Project, University of Alberta, University of Victoria.

Twenty-five per cent enlargement of the portion of no 19 showing the "Ghost"/Angel glacier, Mt Edith Cavell.

Photograph by M.P. Bridgland of Mt Erebus and Eremite Glacier from Thunderbolt Peak, Stn 13, no 102, direction southwest, 1915. Courtesy Library and Archives Canada.

Photograph by J.M. Rhemtulla and E.S. Higgs of Mt Erebus and Eremite Glacier from Thunderbolt Peak, Stn 13, no 102, direction southwest, 1999. Courtesy Rocky Mountain Repeat Photography Project, University of Alberta, University of Victoria.

※

THE RETREAT OF THIS GLACIER during the intervening eighty-five years seems particularly remarkable. Yet, not all pairs of photographs indicate such stark change. Another, from the same station and facing south instead of southwest, offers more evidence of resilience and continuity than of change.

Photograph by M.P. Bridgland of Eremite Mtn and Alcove Mtn from Thunderbolt
Peak, Stn 13, no 97, direction south, 1915. Courtesy Jasper National Park.

Photograph by J.M. Rhemtulla and E.S. Higgs of Eremite Mtn and Alcove Mtn from
Thunderbolt Peak, Stn 13, no 97, direction south, 1999. Courtesy Rocky Mountain
Repeat Photography Project, University of Alberta, University of Victoria.

The Influence on Vegetation Cover
of Fire Suppression and Prevention
in Jasper National Park

———

Photograph by M.P. Bridgland of Athabasca River valley, Grand Trunk Pacific Railway line, and Colin Range
downstream from Jasper town site, Stn 57, no 459, direction north, 1915. Courtesy Jasper National Park.

✳

FROM WHAT EARLIER TRAVELLERS CALLED A SAVANNAH, the valley floor has evolved
into a mixed forest in eighty-five years' time.

Photograph by J.M. Rhemtulla and E.S. Higgs of Athabasca River valley, Grand Trunk Pacific Railway line,
and Colin Range downstream from Jasper town site, Stn 57, no 459, direction north, 1999.
Courtesy Jasper National Park.

Photograph by M.P. Bridgland of Athabasca River valley at Henry House Flats, and Colin Range and Peak,
Stn 58, no 467, direction eastnortheast, 1915. Courtesy Jasper National Park.

※

THE HIGH RESOLUTION of the original and repeat photographs permits detailed comparison of plant communities. In 1915, the Henry House Flats, northeast and downstream from the
town site of Jasper near the modern-day airstrip, supported a more varied grassland community
and woody debris than was found in 1998, when extensive herbivory dominated the same space.
Across the Athabasca River on the lower slopes of the Colin Range, much more extensive tree
cover is evident in the background of the repeat photgraph.

Photograph by J.M. Rhemtulla and E.S. Higgs of Athabasca River valley, Canadian National Railway line, and Colin Range and Peak, Stn 58, no 407, direction eastnortheast, 1998. Courtesy Rocky Mountain Repeat Photography Project, University of Alberta, University of Victoria.

Photograph by M.P. Bridgland of Lac Beauvert, Athabasca River valley, Jasper town site, and Pyramid Mtn from
Old Fort Point, Stn 26, no 220, direction northnorthwest, 1915. Courtesy Library and Archives Canada R214.

✳

THE EXTENSIVE REFORESTATION OF THE VALLEY around the real estate of Jasper Park
Lodge and Jasper town site shows the extensive effect of the policies of fire prevention and fire
suppression that Jasper Park has enforced since 1911. The landscape is utterly transformed.

Photograph by J.M. Rhemtulla and E.S. Higgs of Lac Beauvert, Athabasca River valley, Jasper town site, and Pyramid Mtn from Old Fort Point, Stn 26, no 220, direction northnorthwest, 1998. Courtesy Rocky Mountain Repeat Photography Project, University of Alberta, University of Victoria.

Photograph by M.P. Bridgland of Medicine Lake from Maligne Road, Stn 37, no 296,
direction southeast, 1915. Courtesy Library and Archives Canada R214.

✳

STARK EVIDENCE IN 1915 of the effects of a fire on the hillside extending up from the
shore of Medicine Lake has all but disappeared by 1999. Species such as lodgepole pine depend
on fire for the propagation of its seeds. Frequent natural fires create varied communities of
multiple ages and densities; as well, they reduce the risk of catastrophic fires, such as the type

Photograph by J.M. Rhemtulla and E.S. Higgs of Medicine Lake from Maligne Road,
Stn 37, no 296, direction southeast, 1999. Courtesy Rocky Mountain Repeat
Photography Project, University of Alberta, University of Victoria.

that would result today from an uncontrolled burn in the Athabasca River valley, because the forest cover is so uniformly dense today that any fire would burn at much hotter temperatures than in the past. Prescribed fires, generated and supervised by park staff, are periodically set to begin reducing fuel loads and to create firebreaks. Such management anticipates the return of wildlife without endangering human life and property.

THE INFLUENCE OF PERMANENT HUMAN PRESENCE
IN JASPER NATIONAL PARK

Photograph by M.P. Bridgland of Athabasca River valley at Tri-Valley Confluence, Jasper town site from
the Whistlers, Stn 25, no 208, direction north, 1915. Courtesy Library and Archives Canada R214.

Right: M.P. Bridgland, Stn 25, no 208 as it appeared in his
Description of & Guide to Jasper Park, 14.

Photograph by J.M. Rhemtulla and E.S. Higgs of Athabasca River valley at Tri-Valley Confluence, Jasper town site from the Whistlers, Stn 25, no 208, direction north, 1999. Courtesy Rocky Mountain Repeat Photography Project, University of Alberta, University of Victoria.

✳

THE EXPANSION OF THE TOWN, to the left of the Athabasca River, and the facilities and golf course of Jasper Park Lodge, to the right, are notable, but perhaps no more than the vast changes to the vegetation cover. Consider the forest around Pyramid and Patricia lakes (to the left of the town site and toward Pyramid Mtn). This change can be seen very well from a detail of the photographs of the town site taken from Signal Mtn.

Photograph by M.P. Bridgland of Jasper from Signal Mountain, Stn 27, no 225, direction west, 1915 (detail).
Courtesy Jasper National Park.

Photograph by J.M. Rhemtulla and E.S. Higgs of Jasper from Signal Mountain, Stn 27, no 225, direction west,
1997 (detail). Courtesy Rocky Mountain Repeat Photography Project, University of Alberta,
University of Victoria.

Signs of Transitory Human Presence
in Jasper National Park

Photograph by M.P. Bridgland of Grand Trunk Pacific Railway along Brûlé Lake, Roche Miette in background, Stn 91, no 730, direction southsouthwest, 1915. Courtesy Jasper National Park.

✳

THE FAMOUS SAND DUNES along the northeast shore of this widening of the Athabasca River lay inside the eastern park boundary in 1915, but the redrawing of that boundary in 1929 excluded the lake from the park. In the vicinity of this stretch of rail, which was ripped up in 1917 for the war effort, the Grand Trunk Pacific Railway had its Park Gate station. By 2005, this station has all but disappeared under the shifting sand dunes of the area; only a portion of the roof and gable remain visible.

Photograph by J.M. Rhemtulla, E.S. Higgs, and I.S. MacLaren of Brûlé Lake, Roche Miette in Background, Stn 91, no 730, direction southsouthwest, 1998. Courtesy Rocky Mountain Repeat Photography Project, University of Alberta, University of Victoria.

Details, including inset, of this pair of
photographs reveals the gradual disappearance
of Ewan Moberly's homestead beneath
Mt Esplande.

Photograph by M.P. Bridgland of Athabasca River valley,
mouth of Snaring River, from Mt Esplanade, Stn 62, no 504,
direction southeast, 1915. Courtesy Jasper National Park.

✳

THE ORIGINAL PHOTOGRAPH shows little evidence of the homesteading that occurred in
this stretch of the river valley before 1910, but digitalization reveals the considerable workings of
the homestead of Ewan Moberly and his family, with one of the fields still plainly visible in the
repeat photograph (bisected by the Celestine Lake Road from Snaring River overflow campground). During the last part of the nineteenth century and the first decade of the twentieth,
this and other farmsteads were operated by families residing in the stretch of the Athabasca
River valley between what are now the town site of Jasper and Jasper Lake. The existence of
permanent farms and ranches did not accord with the political and social idea of a wilderness
park when Jasper Forest Reserve was established in 1907. The members of homesteading fami-

Photograph by R.M. Rhemtulla and E.S. Higgs of Athabasca River valley, mouth of Snaring River, from Mt Esplanade, Stn 62, no 504, direction southeast, 1999. Courtesy Rocky Mountain Repeat Photography Project, University of Alberta, University of Victoria.

lies (almost all of them of mixed-blood Iroquois descent) were considered "squatters" on government land. They were modestly compensated for the improvements made to the land and, by virtue of a ban on firearms and hunting, were obliged to relocate to areas outside the park. Indications of the presence of these homesteads haunt this stretch of the upper Athabasca River valley today despite almost a century of designation as a national park. Besides former fields, these include stumps from logging, depressions where farm buildings once stood, burial sites—even the odd rhubarb plant and mushroom patch. Within five years of the homesteaders' departure, Tent City, the precursor to Jasper Park Lodge, was operating on the shores of Lac Beauvert to accommodate tourists attracted to the "wilderness" of Jasper.

Appendices

✳

BRIDGLAND'S ASCENTS

Year   Peaks Climbed (* denotes first ascent)

1902   Selkirks: Ursus Major Mtn*, Ursus Minor Mtn*, Bagheera Mtn*, Rogers Pk, Catamount Pk*, Castor Pk, Bishops Range*, Mt Dawson, Hasler Pk, Mt Donkin, Mt Purity, Wheeler Mt*, Mt Oliver*, Clarke Pk*, Mt Swanzy, Pollux Pk, Mt Abbott, Mt Fox, Mt Selwyn, Beaver Overlook*.

1903   Cdn Rockies: Fairview Mt, Mt Sheol*, Eiffel Pk, Mt Bident, Boom Mt*, Pulpit Pk, Bow Pk, Observation Pk, Cirque Pk, Mt Bosworth*, Mt Niblock, Mt Hector, Mt Gordon, Mt Thompson, Mt Temple, Mt Daly, Mt Aberdeen.

1904   Cdn Rockies: Mt Whymper, Vermillion Pk, Mt Oke*, Misko Mtn*, Mt Owen, Park Mt*, Mt Yukness*, Mt Dennis, Mt Field, Mt Burgess, Michael Pk, Mt Kerr, Mt McArthur, Molar Mtn*, Storm Mt, Mt Niles, Mt Wapta, Mt Balfour, Mt Carnarvon*, Mt Stephen, Mt Des Poilus, Mt King.

1905   Cdn Rockies: Vermillion Pk, Mt Wapta, Kiwetinok Pk, Mt Collie, Mt Duchesnay*, Mt Vaux, Mt Hunter*.

1906   Cdn Rockies: Ptarmigan Pk*, Mt Redoubt*, Fossil Mtn*, Oyster Pk*, Michael Pk, Mt Wapta, The President, The Vice-President, Ogre Pk*, Mt Sealion*, Amiskwi Pk*, Mt Sharp*, Helmet Mtn*, Zinc Mt*.

YEAR   PEAKS CLIMBED (* DENOTES FIRST ASCENT)

Year	Peaks Climbed
1907	Cdn Rockies: Mt Temple, Mt Aberdeen, and survey work in sections where peaks were unnamed.
1908	Selkirks: Rogers Pk, Hermit Mtn, Pollux Pk, Leda Pk.
1909	Cdn Rockies: Mt Huber, Wiwaxy Pk, Mt Burgess, Mt Des Poilus, Mt McArthur.
1910	Selkirks: Mt Mackenzie, Mt Cartier, Mt Sproat, Mt Carnes*. Monashees: Mt Copeland*, Mt Begbie, Griffin Mt, Mabel Mtn.
1911	Cdn Rockies and Selkirks: Cascade Mtn, Storm Mtn, Mt Bonney, N. Albert Pk, Mt Mackenzie, Incompleux Mt, Monashees: Mt Copeland, Mt Begbie, Griffin Mt, Mabel Mtn.
1912	Selkirks: Mt Bonney, N. Albert Pk, Mt Carnes, Cornice Mt, Mt Mackenzie. Monashees: Mt Begbie, Griffin Mt, Mabel Mtn.
1913	Cdn Rockies: Beehive Mt*, Gould Dome*, Pigeon Pk, and minor summits in north part of Crowsnest Forest Reserve.
1914	Cdn Rockies: Cascade Mtn, Storm Mtn, Bonnet Mtn*, Mt Lougheed, Fatigue Mt, Mt McArthur, Mt King, Sofa Mt, and other peaks in south part of the Crowsnest Forest Reserve and Waterton Lakes Park.
1915	Cdn Rockies: Thunderbolt Pk*, Mt Clitheroe*, Mt Maccarib*, Chak Pk*, Aquila Mtn*, Mt Majestic*, Mt Tekarra*, Excelsior Mt*, Roche Bonhomme, Roche De Smet, Roche Miette, Roche Ronde, Mt Aeolus, Mt Utopia.
1918	Cdn Rockies: Mt Prow*, Unnamed 10,189'*, Wapiti Mtn*, other unnamed peaks near Red Deer and Panther rivers.
1919	Cdn Rockies: Wampum Pk*, Mt Malloch, Smoky Mt*, Unnamed 10,189'*, Mt Harris*, other peaks along the Clearwater River.
1920	Cdn Rockies: Bonnet Mtn, Block Mtn*.
1921	Low unnamed peaks in vicinity of Pitt Lake, near Coquitlam, BC.
1922	Cdn Rockies: Mt Verendrye*, other peaks in southern part of Kootenay Park.

Year Peaks Climbed (* denotes first ascent)

1924	Cdn Rockies: Conical Pk*, Unnamed 9379'*, Mt Siffleur*, other peaks along North Saskatchewan River valley to the north.
1927	Cdn Rockies: Mt Wilson, Mt Coleman, Unnamed 10,054'* (southeast of Sunset Pass), Unnamed 10,054'* (north-east of Waterfowl Lakes), Unnamed 9947'* (north of Nigel Pass).
1928	Cdn Rockies: Nigel Pk, Sunwapta Pk, Dalhousie Mt*, Unnamed 9717'* (forks of Brazeau River), Unnamed 10,224'* (southwest of Brazeau River forks), Unnamed 10,264'* (east of Job Pass).

Note: As Bridgland once observed in a conversation with Ley Harris, he stopped counting his first ascents, or, at least, was not always diligent in recording each and every ascent he made. This list provides the names of the mountains he definitely climbed and derives from J. Monroe Thorington, "Members elected after December 1902 and prior to 1 January 1912," *American Alpine Journal* 6 (1947): 345–48.

BRIDGLAND'S SURVEYS

Year	Survey Area
1902	Selkirk Mountains, Beaver River valley, BC
1903	Main Range, Rocky Mountains: Aberdeen Mountain area, BC
1904	Main Range, Rocky Mountains: Baker Peak area, Siffleur Peak area, BC/AB
1905	Main Range, Rocky Mountains: Fall of the Waves, Amisk River Pass, BC
1906	Main Range, Rocky Mountains: BC/AB boundary area
1907	Main Range, Rocky Mountains: Beavermouth—Blaeberry River areas BC/AB
1908	Railway Belt: Shuswap Lake (Ainsley Arm), and Hunawaka Lake areas, north and south of Golden, BC
1909	Railway Belt: Adams Lake area, north and south of Golden, south of Shuswap Lake, BC
1910	Triangulation Surveys in Railway Belt, Revelstoke area, BC
1911	Triangulation Surveys in Railway Belt, Revelstoke area, BC
1912	Triangulation Surveys in Railway Belt, Revelstoke area, BC
1913	Crowsnest Forest Reserve and supervision of Railway Belt surveys
1914	Crowsnest Forest Reserve and supervision of Railway Belt surveys
1915	Jasper Park

Year	Survey Area
1916	
1917	Bow River and Clearwater Forest Reserves: Clearwater River and Red Deer River areas, AB
1918	Bow River and Clearwater Forest Reserves: Red Deer River and Panther River areas, including Dormer River, AB
1919	Bow River and Clearwater Forest Reserves: Red Deer River area, AB
1920	Bow River and Clearwater Forest Reserves: Red Deer River and Ram River areas, AB
1921	Pitt Lake, Port Moody Sheet, BC
1922	South Kootenay Park and Columbia Valley, BC/AB boundary
1923	Kootenay Park and Columbia valley, BC/AB boundary
1924	Clearwater Forest Reserve
1925	Calgary area, AB (incl. aerial work) and Revelstoke district, BC
1926	Gleichen—Drumheller area, AB, and Revelstoke district, BC
1927	Jasper Park area, BC/AB boundary, and Brazeau Lakes area
1928	Jasper Park area, BC/AB boundary
1929	Jasper and Banff Parks' boundaries
1930	Peace River District, BC/AB boundary
1931	Jasper and Banff National Parks' boundaries

Sources

＊

MOST SOURCES CONSULTED are listed alphabetically below. Mention needs to be made of a few general ones for one or another section of this book. An especially significant resource is the *Canadian Alpine Journal*. Dating from 1907, the year after the Alpine Club of Canada was formed, this annual, now in its eighty-eighth volume, contains a wealth of information in the form of scientific essays, historic photographs, and personal mountaineering anecdotes. Aphrodite Karamitsanis's volume of place names in the Alberta mountains, as well as the websites, www.peakfinder.com, maintained by Dave Birrell, and the immense *Canadian Mountain Encyclopedia* http://bivouac.com/default.asp, help confirm place names, locations, and historical details.

Notable online archive indexes and databases include the Archives Network of Alberta, http://asalive.archivesalberta.org:8080/access/asa/archaa and the Glenbow Museum archives and online photograph data bank, www.glenbow.org/collections/search.

ONE

_____

＊ *Measuring the West Before Bridgland* ＊
Exploring, Surveying, and Mapping

BESIDES DON W. THOMSON'S *Men and Meridians*, two useful anecdotal accounts about surveying in western Canada are James MacGregor's *Vision of an Ordered Land*, and Courtney Bond's *Surveyors of Canada*. Early issues of the *Journal of the Dominion Land Surveyors' Association*, later called the *Canadian Surveyor*, and now published under the title *Geomatica*, provide a good idea of the dominant issues concerning surveyors who worked in the 1920s and 1930s.

TWO AND FOUR

✳ *Bridgland's Life and Times 1878–1914* ✳

AND

✳ *Bridgland's Life and Times 1916–1948* ✳

VARIOUS SOURCES WERE CONSULTED, including some referred to above and below. M.P. Bridgland's second son, General Edgar Bridgland, graciously consented to interviews and generously shared his memories of his father with MacLaren and Zezulka-Mailloux. The latter also benefited greatly from similar interviews with Edgar's second cousin, Dr. Margaret B. (Marmie) Hess. For information on the Bridgland and Parsons properties in York, consult *Illustrated Historical Atlas of York County*. The *Dictionary of Canadian Biography*, available electronically at http://www.biographi.ca/EN/ provides detailed information about James W. Bridgland II and his surveys in northern Ontario (the entry for him appears in Volume X of the printed version), and further information concerning the Bridgland and Parsons families is available through the Early Canadiana Online Archives at http://www.canadiana.org/ECO and at the Archives of Ontario. The Victoria University circular, *Acta Victoriana*, offers a most interesting window into turn-of-the-century student life in Toronto, and the two memorials to Bridgland, by C.B. Sissons and E.O. Wheeler, published in the *Canadian Alpine Journal* in 1948, remember Bridgland's character as a friend and colleague.

Much has been written by and about A.O. Wheeler, Bridgland's first climbing and surveying mentor, and two publications in particular have been helpful in piecing together the first few years of Bridgland's career, namely, Esther Fraser's *Wheeler*, and Wheeler's own *The Selkirk Range*.

Correspondence between Bridgland and the office of the surveyor general (Library and Archives Canada, RG 88) was useful in following

the details of the surveyor's career, including the location and particu-
lars of his various surveys. The Department of the Interior's *Annual
Reports* helped determine the chronology of Bridgland's surveys and the
mandates that dictated where and how he conducted his work.

Judy Larmour generously provided anecdotal information
pertaining to later mountaineering work and surveys, and pointed out
the existence of Lizzie Rummel's second recorded interview with Ley
Harris, Bridgland's steady assistant in the field for many years.
Throughout the three interviews that eventually came to light, Harris
talks of his "experiences and incidents" with Bridgland, and introduces
many of his anecdotes with the phrase, "well, there was a time that Mr
Bridgland told me of...."

Bridgland's "Report of the Chief Mountaineer," which he published
annually until 1910 in the *Canadian Alpine Journal*, lists graduating
members, their guides, and the routes they used, as well as the conditions they encountered. To this information, several of Harris' stories
were added, as were some of the stories related by C.B. Sissons in his
book of memoirs.

THREE

———

✳ *Bridgland's Survey of Jasper Park 1915* ✳

IN MOST FIELD SEASONS Bridgland kept a diary in a hard-cover
notebook issued by the Department of the Interior. Comprising meteorological details and short notes listing the stations occupied each
day, the one for 1915 provided the context for Section I of this book.
Jeanine Rhemtulla's notes on the modern names for both camera
stations and peaks helped clarify Bridgland's diary entries, made when
many peaks in the area either were unnamed or bore different names
from those they have today. Several sources helped to flesh out the 1915
survey and its historical landscape. Chief among these is the correspondence between Bridgland and Surveyor General Édouard-Gaston
Deville, held in the LAC's "Surveys and Mapping" record group, RG
88. These discuss all aspects of the season's work: letters from Deville
give Bridgland his mandate for the summer; memos sent by Bridgland

list the materials he needed; and periodic reports detail the survey's progress and problems. Secondly, Bridgland and Douglas's *Description of & Guide to Jasper Park* contains accounts of some of the trails he used, the flora and fauna he identified, and the landscapes he surveyed. Many more details than the present book includes provide useful information about the routes used to access the camera stations. Thirdly, http://parkscanada.pch.gc.ca/pn-np/ab/jasper/index_e.asp, the official website of Jasper National Park, offers facts about the history, and it lists photographic and written histories of the park.

FIVE

※ *Bridgland's Legacy* ※

*Culture, Ecology, and Restoration in Jasper National Park,
and the Rocky Mountain Repeat Photography Project*

PRIMARY SOURCES FOR THIS SECTION also included interviews with Jeanine Rhemtulla and Eric Higgs in 2002, 2003, and 2004. With enthusiasm, both recounted to MacLaren or Zezulka-Mailloux the adventures and misadventures of their rephotography of Bridgland's survey. The RMRPP website served as a useful reference guide for this and the other sections. (All titles mentioned above are detailed below.) *Vantage Points* and *The Rockies through the Lens of Time*, the two exhibitions curated by Trudi Smith, provided good public response to Bridgland's work and feedback on the repeat photography project's methodology and the best ways of presenting its findings. ※

Alford, Ethelwyn Octavia Doble. "With the Forest Rangers in the Rockies: 2 July 1924-5 October 1924." Unpublished Diary, 1924. (Photocopy of MS held by University of Alberta Library, Edmonton.)

Barnett, Douglas E. "The Deville Era: Survey of the Western Interior of Canada." *Alberta History* 48 (Spring 2000): 19–25.

Beck, Janice Sanford. *No Ordinary Woman: The Story of Mary Schäffer Warren.* Calgary: Rocky Mountain Books, 2001.

Benham, D.J. "Jasper Park in the Canadian Rockies: Canada's New National Playground." *The Globe*, Saturday Magazine section, 15 Jan. 1910. 4, 9.

Berger, Paul. "Doubling: This Then That." *Second View: The Rephotographic Survey Project.* Ed. Mark Klett, Ellen Manchester, and JoAnn Verburg. Albuquerque: University of New Mexico Press, 1984. 45–52.

Birrell, A[ndrew] J. "Classic Survey Photos of the Early West." *Canadian Geographical Journal* 91.4 (Oct. 1975): 12–19.

———. "Fortunes of a Misfit; Charles Horetzky." *Alberta Historical Review* 19 (Winter 1971): 9–25.

———. *Into the Silent Land: Survey Photography in the Canadian West, 1858-1900.* Ottawa: Public Archives of Canada, 1975.

———. "The North American Boundary Commission: Three Photographic Expeditions, 1872-74." *History of Photography* 20 (1996): 113–21.

———. "Survey Photography in British Columbia, 1858-1900." *BC Studies* 52 (Winter 1981–82): 39–60.

Birrell, Dave. *50 Roadside Panoramas of the Canadian Rockies.* Calgary: Rocky Mountain Books, 2000.

Bond, Courtney C.J. *Surveyors of Canada, 1867-1967.* Ottawa: Canadian Institute of Surveying, 1966.

Bridgland, M[orrison] P[arsons]. "Electrical Discharge in Air at Low Pressure," *Acta Victoriana* 24.4 (1901): 361–64.

———. "Jasper Park." *Canadian Alpine Journal* 10 (1919): 70–77.

———. *Photographic Surveying.* Dept. of the Interior Bulletin 56. Ottawa: F.A. Acland, 1924.

———. "Report of the Chief Mountaineer." *Canadian Alpine Journal* 1 (1907): 171–75.

———. "Report of the Chief Mountaineer." *Canadian Alpine Journal* 1 (1908): 329–32.

———. "Report of the Chief Mountaineer." *Canadian Alpine Journal* 2 (1909): 154–57.

———. "Some Trails between Banff and Nordegg." *Canadian Alpine Journal* 18 (1930): 50–59.

———, and A.J. Campbell. "Notes by M.P. Bridgland, D.L.S., and A.J. Campbell, D.L.S." In A.O. Wheeler. "The Application." 87–96.

———, and Robert Douglas. *Description of & Guide to Jasper Park.* Ottawa: Dept. of the Interior, 1917.

Brown, R.C. "The Doctrine of Usefulness: Natural Resource and National Park Policy in Canada, 1887–1914." *The Canadian National Parks: Today and Tomorrow.* Eds. J.G. Nelson and R.C. Scace. Calgary: University of Calgary, 1968; Montreal: Harvest House, 1969. 46–62.

Canada, Department of the Interior. *Map showing Areas which it is Proposed to withdraw from Rocky Mountain and Jasper Parks* 1:792,000. Ottawa: Dept. of the Interior, 1929. (Copy held by the Park Library, Jasper National Park, Jasper.)

———. *A Sprig of Mountain Heather: Being a Story of the Heather and some Facts about the Mountain Playgrounds.* Ottawa: Dept. of the Interior, 1914.

———, Topographical Surveys Branch. *General Instructions to Surveyors in charge of Parties for the Survey of Dominion Lands in Manitoba, Saskatchewan, Alberta, the Northwest Territories and the Block of Three and One-Half Million Acres in British Columbia.* Ottawa: Department of the Interior, 1913. (Copy in LAC, RG 88, vol. 140, file 13446.)

———, Statutes, 20–21 George V, Chap. 33. *"The National Parks Act."* Assented to 30 May 1930. Rptd. in *Documenting Canada: A History of Modern Canada in Documents.* Eds. Dave De Brou and Bill Waiser. Saskatoon: Fifth House, 1992. 299–302.

———, Statutes, 50–51 Victoria, Chap. 32. "Rocky Mountains Park Act, 1887." Assented to 23 June 1887. Rptd. in *Documenting Canada: A History of Modern Canada in Documents.* Eds. Dave De Brou and Bill Waiser. Saskatoon: Fifth House, 1992. 154–55.

Cavell, Edward, and Jon Whyte. *Rocky Mountain Madness: A Bittersweet Romance.* Banff: Altitude Publishing, 1982.

Davis, Raymond E., Francis S. Foote, and Joe W. Kelly. *Surveying: Theory and Practice.* 1928. 5th ed. New York: McGraw-Hill, 1966.

Deville, Édouard-Gaston. *Photographic Surveying, including the Elements of Descriptive Geometry and Perspective.* Ottawa: Government Printing Bureau, 1889. 2d ed. Ottawa: Survey Office, 1895.

Dolgoy, R. Interview with L.E. Harris, 13 Feb. 1978. Phonotape, Provincial Museum of Alberta, Edmonton, 78.65/17.

Doyle, Arthur Conan. *Memories and Adventures.* London: Hodder and Stoughton, 1924.

———. "Western Wanderings." *Cornhill Magazine,* 112/38, 3d ser. (Jan.-Apr. 1915): 1–12, 145–52, 289–97, 433–43.

———. *Western Wanderings.* Ed. and introd. Christopher Roden and Barbara Roden. Penyffordd, Eng.: Arthur Conan Doyle Society, 1994.

Fischer, Ron. "Seeing with Metric Eyes: the Unknown Origins of Motion Capture, Part 1." *VR News* 9:2 (Mar. 2000).

Fraser, Esther. *The Canadian Rockies: Early Travels and Explorations.* Edmonton: Hurtig, 1969.

———. *Wheeler.* Banff: Summerthought, 1978.

Friesen, Gerald. *The Canadian Prairies: A History.* Toronto: University of Toronto Press, 1987.

Gainer, Brenda. "The Human History of Jasper National Park." Report Ser.
no. 441. Typescript Report for Parks Canada, 1981. (Copy held by the Park
Library, Jasper National Park, Jasper.)

Goetzmann, William H. *Exploration and Empire: The Explorer and the Scientist in the
Winning of the American West.* 1966. New York: Norton, 1978.

Grant, George Monro. *Ocean to Ocean: Sandford Fleming's Expedition through Canada
in 1872. Being a Diary kept during a Journey from the Atlantic to the Pacific with the
Expedition of the Engineer-in-Chief of the Canadian Pacific and Intercolonial Railways.*
Toronto: James Campbell and Son; London: Sampson, Low, Marston, Low,
and Searle, 1873.

Great Plains Research Consultants. "Jasper National Park: A Social and
Economic History." Typescript Report for Parks Canada, 1985. (Copy held
by the Park Library, Jasper National Park, Jasper.)

Green, Ashdown H. "Diary of Ashdown H. Green (transit man) C.P.R. Survey
Party 'S' Jan. 31, 1872–Oct. 28, 1873." Transcr. Winnifreda Macintosh, 1966.
BCARS, MS-0437.

Hart, E.J. *Diamond Hitch: The Early Outfitters and Guides of Banff and Jasper.* Banff:
Summerthought, 1979.

———. *The Selling of Canada: The CPR and the Beginning of Canadian Tourism.* Banff:
Altitude Publishing, 1983.

———, ed. *A Hunter of Peace: Mary T.S. Schäffer's Old Indian Trails of the Canadian
Rockies: incidents of camp and trail life, covering two years' exploration through the Rocky
Mountains of Canada / with photographs by the author and by Mary W. Adams and others;
1911 Expedition to Maligne Lake and Yahe-Weh. . . .* Banff: Whyte Museum of the
Canadian Rockies, 1980.

Hastings, James R., and Raymond M. Turner. *The Changing Mile: An Ecological Study
of Vegetation Change with Time in the Lower Mile of an Arid and Semiarid Region.*
Tucson: University of Arizona Press, 1965.

Henderson, Gavin. "James Bernard Harkin: The Father of Canadian National
Parks." *Borealis* 5:2 (Fall 1994): 28–33.

Higgs, Eric. *Nature by Design: People, Natural Process, and Ecological Restoration.*
Cambridge, Mass.: MIT Press, 2003.

———, et al., *Culture, Ecology and Restoration in Jasper National Park: An Interdisciplinary
Research Project.* Dept. of Anthropology, University of Alberta, 1999.

"Historical Sketch of Topographic Surveys Branch, Prepared in 1924 for the
Natural Resources Intelligence Branch, under the direction of Thomas
Shanks, Assistant Director General of Surveys." LAC, RG 88, vol 405, file
23/262.883-9 pt.1.

Huyda, Robert J. *Camera in the Interior, 1858: H.L. Hime, Photographer: The Assiniboine and Saskatchewan Exploring Expedition.* Toronto: Coach House, 1975.

Illustrated Historical Atlas of York County, Ontario. Toronto: Miles and Co., 1878; Rpt. Belleville, Ont.: Mika Silk Screening, 1972.

Jenish, D'Arcy. *Epic Wanderer: David Thompson and the Mapping of the Canadian West.* Toronto: Doubleday Canada, 2003.

Kain, Conrad. Kain, *Where the Clouds Can Go.* Ed. and supplemented J. Monroe Thorington. New York: American Alpine Club, 1935.

Karamitsanis, Aphrodite. *Place Names of Alberta, Vol. 1: Mountains, Mountain Parks and Foothills.* Edmonton: Alberta Community Development, Friends of Geographical Names of Alberta Society, University of Calgary Press, 1991.

Larmour, Judy. *Laying Down the Lines: A History of Land Surveying in Alberta.* Edmonton: Brindle & Glass, 2005.

Lewis, Macolm G., ed. *Cartographic Encounters: Perspectives on Native American Map-making and Map Use.* Chicago: University of Chicago Press, 1998.

Linklater, Andro. *Measuring America: How an Untamed Wilderness Shaped the United States and Fulfilled the Promise of Democracy.* New York: Walker and Co., 2002.

Lothian, W.F. *A Brief History of Canada's National Parks.* Ottawa: Minister of the Environment, Minister of Supply and Services Canada, 1987.

Luxton, Eleanor G. *Banff, Canada's First National Park: A History and a Memory of Rocky Mountains Park.* Banff: Summerthought, 1975.

McEvoy, James. *Report on the Geology and Natural Resources of the Country traversed by the Yellow Head Pass Route from Edmonton to Tête Jaune Cache, comprising Portions of Alberta and British Columbia.* Geological Survey of Canada, Annual Report, vol. 11, pt D. Ottawa: S.E. Dawson, 1900.

McFetridge, Robert. "The Alberta Connection: Conan Doyle in Alberta—1914" http://bakerstreetdozen.com/mcfetridge.html.

McGinnies, William J., Homer L. Shantz, and William G. McGinnies. *Changes in Vegetation and Land Use in Eastern Colorado.* Washington, D.C.: U.S. Department of Agriculture [ARS-85], 1991.

MacGregor, J.G. *Overland by the Yellowhead.* Saskatoon: Western Producer Prairie Books, 1974.

———. *Vision of an Ordered Land: The Story of the Dominion Land Survey.* Saskatoon: Western Producer Prairie Books, 1981.

MacLaren, I.S. "Cultured Wilderness in Jasper National Park." *Journal of Canadian Studies* 34.3 (Fall 1999): 7–58. Also available electronically at http://bridgland.sunsite.ualberta.ca/jasper/publications.html

Marty, Sid. *A Grand and Fabulous Notion: The First Century of Canada's Parks.* Toronto: NCP, 1984.

Meagher, Mary, and Douglas B. Houston, *Yellowstone and the Biology of Time: Photographs Across a Century*. Norman: University of Oklahoma Press, 1998.

Moberly, Walter. "Canadian Pacific Railway Exploratory Survey, Diary and Notes of Walter Moberly Engineer in Charge of Party S, April 19, 1872–February 16, 1873." Transcr. Fred Howlett, 1998. BCARS, file 334–16–12.

"Morrison Parsons Bridgland." *The Canadian Surveyor*, 9:7 (Jan. 1948): 21.

Murray, Jeffrey S. "Mapping the Mountains." *Legion Magazine* 76:3 (May/June 2001). 47. Also available electronically at http://www.legionmagazine.com/features/canadianreflections/01-05.asp

Oltman, Ruth. *Lizzie Rummel: Baroness of the Canadian Rockies*. Exshaw, Alberta: Ribbon Creek Publishing, 1983.

Palliser, John, *et al. The Journals, detailed Reports, and Observations relative to the Exploration, by Captain Palliser, of that Portion of British North America, which, in Latitude, lies between the British Boundary Line and the Height of Land or Watershed of the northern or frozen Ocean respectively, and in Longitude, between the western shore of Lake Superior and the Pacific Ocean during the years 1857, 1858, 1859, and 1860; presented to both Houses of Parliament by command of Her Majesty, 19th May, 1863*. London: Printed by G.E. Eyre & W. Spottiswoode, 1863.

Parker, Elizabeth. "The Alpine Club of Canada," *Canadian Alpine Journal* 1 (1907): 3.

————, and A.O. Wheeler. (See Wheeler.)

Pickard, John. "Assessing Vegetation Change over a Century using Repeat Photography." *Australian Journal of Botany* 50 (2002): 409–14

Reichwein, PearlAnn. "Parker, Elizabeth." *Biographical Dictionary of American and Canadian Naturalists and Environmentalists*. Ed. Keir B. Sterling, *et al.* Westport, Conn.: Greenwood Press, 1997, 609–10.

————. "At the Foot of the Mountain: Preliminary Thoughts on the Alpine Club of Canada, 1906-1950." *Changing Parks: The History, Future and Cultural Context of Parks and Heritage Landscapes*. Ed. John S. Marsh and Bruce W. Hodgins. Toronto: Natural Heritage, 1998. 160–76.

————, and Karen Fox. "Margaret Fleming and the Alpine Club of Canada: A Woman's Place in Mountain Leisure and Literature, 1932-1952." *Journal of Canadian Studies* 36.3 (Fall 2001): 35–60.

————, eds. *Mountain Diaries: The Alpine Adventures of Margaret Fleming 1929-1980*. Edmonton: Historical Society of Alberta, 2004.

Reid, Dennis. "Photographs by Tom Thomson." National Gallery of Canada, *Bulletin*, no. 16 (1970): 2–36. Also available electronically at http://collections.ic.gc.ca/bulletin/num16/reid1.html

Reithmaier, Tina. "Maps and Photographs." *The Historical Ecology Handbook: A Restorationist's Guide to Reference Ecosystems*. Eds. D. Egan and E.A. Howell. Washington, D.C.: Island Press, 2001. 121–46.

Rhemtulla, Jeanine. "Eighty Years of Change: The Montane Vegetation of Jasper
    National Park." M.Sc. Thesis. Department of Renewable Resources,
    University of Alberta, 1999.

——, et al. "Eighty Years of Change: Vegetation in the Montane Ecoregion
    of Jasper National Park, Alberta, Canada," *Canadian Journal of Forest Research* 32
    (2002): 2010–21.

Roberts, A.C. "The Surveys in the Red River Settlement in 1869." Supplement
    to *Canadian Surveyor* 24 (June 1970): 238–48.

Robson, Joan. "The Glacier Trail: Jasper to Lake Louise, July 3 to 23, 1927."
    Unpublished paper, 1952. (Copy courtesy Tom Peterson, Hinton, Alberta.)

Rocky Mountain Repeat Photography Project (RMRPP).
    http://bridgland.sunsite.ualberta.ca/main/index.html

Rogers, Garry F., Harold E. Malde, and Raymond M. Turner. *Bibliography of Repeat
    Photography for Evaluating Landscape Change.* Salt Lake City: University of Utah
    Press, 1984.

Rose, Gillian. *Visual Methodologies: An Introduction to the Interpretation of Visual Materials.*
    Thousand Oaks, Calif.: Sage Publications, 2001.

Rummell, Lizzie. Interviews with Ley Edward Harris, 4 Sept. 1972, 16 Mar. 1976;
    Elizabeth Rummel Fonds, Whyte Museum of the Canadian Rockies,
    M28/S9/V554. (Cassette copies held by the Association of Alberta Land
    Surveyors, Edmonton.)

Rylatt, R.M. *Surveying the Canadian Pacific: Memoir of a Railroad Pioneer.* Fwd. William
    Kittredge. Salt Lake City: University of Utah Press, 1991.

Sandford, Robert W. *The History of Mountaineering in Canada.* Vol 1. *The Canadian
    Alps.* Banff: Altitude Publishing, 1990.

Scott, Chic. *Pushing the Limits: The Story of Canadian Mountaineering.* Calgary: Rocky
    Mountain Books, 2000.

Service, Robert W. *The Spell of the Yukon.* New York: Barse and Hopkins, 1907.

Silversides, Brock V. *Waiting for the Light: Early Mountain Photography in British Columbia
    and Alberta, 1865-1939.* Saskatoon: Fifth House, 1995.

Simpson, Æmilius. "Journal of a Voyage across the Continent of North
    America in 1826." HBCA, PAM B.223/a/3, 50 fols.

"Sir A. Conan Doyle Sails for Canada." *The Globe* 21 May 1914, 1.

Sissons, C.B. "In Memoriam: M.P. Bridgland." *Canadian Alpine Journal* 31 (1948):
    218–20.

——. *Nil Alienum: The Memoirs of C.B. Sissons.* Toronto: University of Toronto
    Press, 1964.

Sleigh, Daphne. *Walter Moberly and the Northwest Passage by Rail.* Surrey, BC: Hancock
    House, 2003.

Smith, Cyndi. *Jasper Park Lodge: In the Heart of the Canadian Rockies*. Canmore, Alberta: Coyote Books, 1985.

Smith, Trudi. "Vantage Points: Scientific Photography in Jasper National Park." MA Thesis, Interdisciplinary Studies, University of Victoria. 2004.

———, curator. *The Rockies through the Lens of Time: The Work of Morrison Parsons Bridgland and the Rocky Mountains Repeat Photography Project*. Main Foyer, Library and Archives Canada, Ottawa, 26 Apr. 2005–31 Jan. 2006.

———, curator. *Vantage Points: Scientific Photography in Jasper National Park; The Bridgland & Repeat Photography Projects*, Showcase Gallery, Jasper-Yellowhead Museum and Archives, Jasper, 5 July–26 Oct. 2003. Thereafter, archived at http://bridgland.sunsite.ualberta.ca/display/.

Spry, Irene, ed. *The Papers of the Palliser Expedition, 1857-1860*. Publications of the Champlain Society, vol. 44. Toronto: Champlain Society, 1968.

Stewart, Lillian. "Burgess Shale Site." *Canadian Encyclopedia*, 2d ed., 4 vols. Edmonton: Hurtig, 1988. 299.

Stutfield, Hugh M., and J. Norman Collie. *Climbs and Exploration in the Canadian Rockies*. New York: Longmans, Green, 1903.

Talbot, F.A. *The New Garden of Canada: By Pack-Horse and Canoe through Undeveloped New British Columbia*. London: Cassells, 1911.

Taylor, C.J. "Legislating Nature: The National Parks Act of 1930." *Canadian Issues/Thèmes canadiens, vol XIII, 1991: To See Ourselves/To Save Ourselves: Ecology and Culture in Canada*. Eds. Rowland Lorimer, Michael M'Gonigle, Jean-Pierre Revéret, and Sally Ross. Montreal: Association for Canadian Studies, 1991. 125–37.

Thomson, Don W. *Men and Meridians*. 3 vols. Ottawa: Queen's Printer, 1966–69. *Volume 2: The History of Surveying and Mapping in Canada 1867-1917*. 1967.

Thorington, J. Monroe. *A Climber's Guide to the Rocky Mountains of Canada*. New York: American Alpine Club, 1921.

———. *The Glittering Mountains of Canada: A Record of Exploration and Pioneer Ascents in the Canadian Rockies, 1914-1924*. Philadelphia: J.W. Lea, 1925.

———. "Members elected after December 1902 and prior to 1 January 1912." *American Alpine Journal* 6 (1947): 345–48.

Veblen, Thomas T., and Diane C. Lorenz. *The Colorado Front Range: A Century of Ecological Change*. Salt Lake City: University of Utah Press, 1991.

Waiser, W.A. *Park Prisoners: The Untold Story of Western Canada's National Parks, 1915-1946*. Saskatoon: Fifth House, 1995.

Washburn, Stanley. *Trails, Trappers, and Tender-feet in the new Empire of Western Canada*. London: A. Melrose, 1912.

Wates, Cyril G. "Annual Report of the Photographic Committee." *Canadian Alpine Journal* 10 (1919): 114–18.

Webb, Jenaya. "Imaging and Imagining: Mapping, Repeat Photography and Ecological Restoration in Jasper National Park." MA Thesis, Department of Anthropology, University of Alberta, 2003.

Webb, Robert H. *Grand Canyon, a Century of Change. Rephotography of the 1889-1890 Stanton Expedition.* Tucson: University of Arizona Press, 1996.

Wheeler, A.O. "The Application of Photography to the Mapping of the Canadian Rocky Mountains." Report submitted to the Alpine Congress at Monaco by the Alpine Club of Canada. *Canadian Alpine Journal* 11 (1920): 76–96.

————. *The Selkirk Range,* 2 vols. Ottawa: Government Printing Bureau, for the Department of the Interior, 1905.

————, [and Elizabeth Parker]. *The Selkirk Mountains: A Guide for Mountain Climbers and Pilgrims.* Winnipeg: Stovel, 1912.

Wheeler, E.O. "In Memoriam." *Canadian Alpine Journal* 31 (1948): 220–22.

"Where the Heather Blooms in Canada. Novel Means of Advertising the Dominion's National Parks." *The Globe* 14 Aug. 1914. 6.

Whittaker, John A., comp. and ed. *Early Land Surveyors of British Columbia (P.L.S. Group).* Victoria, B.C.: Corporation of Land Surveyors of the Province of British Columbia, 1990.

Wilford, *The Map-makers.* 1981. Rev. ed. New York: Alfred A. Knopf, 2000.

Williams, Mabel, ed. *The History and Meaning of the National Parks of Canada: Extracts from the Papers of the late Jas. B. Harkin, first Commissioner of the National Parks of Canada.* Saskatoon: H.R. Lawson Publishing, 1957.

Zeller, Suzanne. *Inventing Canada: Early Victorian Science and the Idea of a Transcontinental Nation.* Toronto: University of Toronto Press, 1987.

Zezulka-Mailloux, Gabrielle. *The Nature of the Problem: Wilderness Paradoxes in Jasper National Park.* PhD. Diss. Department of English, University of Alberta, 2005.

Index

✳

PAGE NUMBERS in bold font indicate where the name of a person (besides Bridgland), place, or geographical feature occurs in the caption to a photograph. Page numbers in italics indicate where the name of a place or geographical feature appears on a map. Titles of publications and of legislation that have been indexed appear under the name of their authors or sponsoring organizations. ✳

Abbot Glacier 187
Abbot Pass 187
Abbott, Mt 56, 61
Aberdeen, Mt 57, 77
Aberhart, William 201
Abraham Lake 92, 182
Aeolus, Mt 219
Akamina Pass 85–86, 90
Alaska Boundary Dispute 23
Alberta Land Surveyors' Association
    XIV, 195
Alcove Mtn 113, **247**
Aldous, Montague 15
Alford, Bert 191
Alford, Ethelwyn Octavia Doble 191–92
Alpine Club of Canada 67–69, 96, 127,
    172, 175–76, 181, 187, 199–200,
    202, 214, 271
    The Alpine Herald 82
    Canadian Alpine Journal 27, 69, 75–76,
      212–13, 271–73
    first ACC club house, Banff **77**, 78
    first annual camp, 1906 **74**
    founders **70**
    founding of, 1906 70–71
    photographic committee 212
Alpine Club (England) 69, 174
American Alpine Club 69
    American Alpine Journal 193
Amethyst Lakes 113, 128, **129**, 130, **141**,
    143, **166**, **178**, **204**
Amgadaom Point 76
Angel Glacier, Mt Edith Cavell. *See*
    Ghost Glacier
Angus, Richard Bladworth 21
Annette, Lake 115, **116**
Appalachian Mountain Club 69
Aquila, Mt 127, 143
Arago, Dominique François Jean 25
Arches National Park (USA) 201
Arris, Mt 113, 130
Arrowsmith, Aaron 7
Ashlar Ridge **135**
Assiniboine, Mt 64
Assiniboine Portage 7
Association of Dominion Land
    Surveyors 195, 200

Astoria River *113*, 126, 128, 130, 173–75
Asulkan Creek *56*, 60
Asulkan Glacier *56*, 61, 63
Athabasca Falls 125
Athabasca Gorge **203**
Athabasca Pass XI, 4, **6**, 7–8, *113*, 130,
          174, 176
Athabasca River and Valley 4, *5*, 6,
          9–10, 16–22, 29–30, **31**, 33,
          35, 101–03, 106, 109–18, *113*,
          116–17, 121–23, **122**, 125–27,
          131–34, **134**, 136–38, **137**, 143,
          149, 153, 155, 173, 175–76,
          177, 215, **234–35**, **248–51**,
          252–53, 256–57, **262–63**
Avalanche, Mt *56*, 78

Bailey, Trish 234
Banff, Alberta *57*, 80, 84, 111, 199, 214
Banff Hot Springs Reserve
          (Cave and Basin) 23
Banff-Jasper Highway 103, 195
Banff National Park (after 1930)
          196–98, 208
Banff Park. *See* Rocky Mountains Park
Barbican Peak (The Ramparts) **168**, **178**
Barkerville, British Columbia 96
base lines (grid survey) 13, 15, 29, 38,
          40–42, 136
Bastion Peak (The Ramparts) **168**, **178**
Bear Creek *56*, 61
Beautemps-Beaupré, Charles François 25
Beauvert, Lac (Horseshoe Lake) 111, *113*,
          116, 123, 153, 209, 218, **252–53**
Beaver Mines, Alberta 85, 87
Beaver River *56*, 59
Beaverfoot River 29
Beavermouth, British Columbia 67
Bedson, Alberta *113*, 136–37, 153
Bedson Ridge 30, 110, *113*, **134**
Begbie, Mt *56*, 83–84
Bell, John 219, 228, 240–41
Belly River 15
Benham, D.J. 103
Benner, Pete 195
Bennett, R.B. 196

Benoit, Pete 195
Berger, Paul 146
Black Cat Mtn 110, *113*, 133, 150
Blackhorn, Mt *113*, 143
Blaeberry River *56*, 67, 95
Blakiston Creek 86, **89**
Blakiston, Mt 86, **89**
Boom, Mtn *57*, 65
Bosworth, Mt *57*, 65
Bow Range (mtns) *57*, 77
Bow River 5, *6*, 24, 27, 29, 36, *57*, 80,
          96, 194
Bow River Forest Reserve *5*, 84, 85, 181,
          186, **188–89**, 204
Brazeau River *5*, 68, *182*, 209
Brewster, Jack 195
Brewster (Ya-Ha-Tinda) Ranch *182*, 208
Bridgland, Charles 67, 81, 139, 201
Bridgland, Clarke 48
Bridgland, Edgar X, XIII, 67, 81, 198,
          201–02, 233, 272
Bridgland, Eleanor Beaton 45
Bridgland, Elizabeth (*née* Johnson) 48
Bridgland family farm (Downsview,
          Ontario) **47**, 48–49, 91, 201
Bridgland Glacier *56*, 206
Bridgland, Hannah Matilda
          (*née* Parsons) 48
Bridgland, James William 45–46
Bridgland, James William II 46–48, 272
Bridgland, James William III 48
Bridgland, Marie (*née* Dennis) 48
Bridgland, Martha Ann (*née* Jones) 48
Bridgland, Mary Elizabeth (*née* Perkins)
          67, 79–80, 139, 201, 204
     character 67–68
Bridgland, Morrison Parsons
          (*né* Alfred Morrison Parsons)
          (1878–1948) XI–XIV, 2, 4, *5*, 10,
          18, 21, 24, 29–30, **35**, 36, 48, 58,
          63, 72–73, 75, 79, 90, 96, 176,
          181, 204, 216–20, 222–23, 226,
          229–37, 239–40, 244, 265–69,
          272–74
     adoption by James and Hannah
          Bridgland 49

Alpine Club of Canada involvement
    appointed head guide/chief
        mountaineer at ACC
        annual camp 69
    awarded ACC's Silver Rope 200
    elected co-vice-president of
        ACC 79
    elected to committee organizing
        first ACC annual camp 71, 73
    photographed at first ACC
        camp, 1906 **74**
    photographed with ACC camp
        guides, 1906 **75**
    photographed with other
        co-founders of ACC **70**
    responses to as a guide 81–82
    role in ACC contrasted to
        Wheeler's 72
    wins ACC photography
        contest 201
birth to William and Annie Parsons
    48
Brazeau River his favourite region
    209
career as DLS
    becomes an author 169–81
    *Description of & Guide to Jasper
        Park* (1917) XI, 39, 59, **110**, 154,
        168, 169–81, **171**, **177–79**, 213, 274
    career ascents 100, 265–67
    career terminated 196–200
    DLS limited examination
        passed 53, 93
    hired as assistant to DLS 50
    Jasper Park survey 99–138, 215
        Jasper Park survey's
        achievement 148
    map-making technique
        described/compared 100,
        139–45, 237
    *Map of the Central Part of Jasper
        Park, Alberta* (1916) 143, **144**,
        160–66, 169
    *Map of Crowsnest Forest Reserve and
        Waterton Lakes Park* (1915) **89**,
        198–99

Photographic Surveying (1924) 107,
    112, 153, 192, 218
photographic technique 117–20,
    124
photographs developed in the
    field 123, 183
primary surveyor of Railway
    Belt surveys 83
promoted to DLS 66
promoted to Surveys Engineer,
    Grade 4 191
*Report of the Triangulation of the
    Railway Belt of British
    Columbia Between Kootenay
    and Salmon Arm Bases* (1915) 97
salary a source of
    discontentment 193
character
    as compared to A.O. Wheeler 55
    inclined toward athletics and
        photography 49–50
    inclined towards teaching 50,
        100, 116
    shyness 237
    understatedness 60, 76
education 49–51
life after the DLS 200–204
marries Mary Elizabeth Perkins
    (1908) 67
nicknamed "Bridge" by Wheeler 78
particular alpine interests (glaciers,
    wildflowers) 213, 237
photographed with siblings **67**
    (1905), **205** (1945)
survey photographs as art 145–48
Bridgland, Mt VIII, *113*, 206
Bridgland Pass *56*, 206
Bridgland Peak *56*, 206
Brinkman, Jack 185–86
British Columbia (terms of entry into
    Confederation) 12, 50
British Mtns (Brooks Range) 23
Brooks Range (British Mtns) 23
Broughton, William 7
Brown, Ernest Percy 31
Brownlee, John Edward 196

Brule, Alberta 30, *113*
Brûlé Hill 150
Brûlé Lake 9, 19, 30, *113*, *133*, 213, **260–61**
Burgess, Mt *57*, 76
Burgess Pass *57*, 66
Burgess Shale 66, 205

Cairngorm, Mt *113*, 123, 143
Calgary, Alberta *5*, 54, 64, 68–69, 77,
      85–87, 96, 100, 106–07, 132,
      138–40, 191–92, 194–95, 199,
      201, 205
camera lucida 26
Cameron Bend (Buchanan Ridge)
      86, **89**
Cameron Creek/Brook 86, **89**
Cameron Lake 86
Cameron Ridge 86
Campbell, Dan **70**
Campbell, R. **70**
Campbell, Robert H. 102–03
Campbell, Sandy 235
Canada
    Act of Union (1841) 91
    Canadian Pacific Railway Act (1881) 21
    Dominion Forest Reserves Act (1906) 150
    Dominion Forest Reserves and Parks Act
      (1911) 102
    Dominion Lands Act (1872) 13
    National Parks Act (1930) 42
    North-West Irrigation Act (1894) 23
    Order in Council 1907–1323 (1907)
      (creates Jasper Forest Park)
      102
    Rocky Mountains Park Act (1887) 28
Canadian Geographical Board 35
Canadian Irrigation Survey 23, 25
Canadian National Railways 91, 150–51,
    226
Canadian Northern Railway (CNoR)
    21, 101, 136, 150–51
CNoR and GTPR lines in
    Jasper Park compared 151
CNoR stations between Hinton
    and Jasper 151
history of completion 149

Canadian Pacific Railway (CPR) 21–22,
    54–55, 84, 87, 107, 201.
    See also Railway Belt
Canmore, Alberta 96, 156
Cardston, Alberta 199
Carlyle, Thomas 51, 91
Carman, Bliss 59
Carmichael, Franklin 147
Carthew, Mt **89**
Cartier, Jacques 2
Castle River 85
Cataract Creek 185
Cautley, Richard W. 94
Cave and Basin Springs (Banff) 23, 36
Cavell Creek *113*, 173
Cavell, Edith 130; 173
Cavell Lake 173–74
Cavell Meadows **177**, **244–45**
Cavell, Mt Edith. *See* La Montagne
C.B. Parsons Middle School (Toronto)
    202
Celestine Lake Road (JNP) 226
CER Project. *See* culture
Chak Peak *113*, 127
Chetamon, Mt *113*, 144
choss heaps **225**
Chrome Lake *113*, 200
Circus Valley *113*, 126, **129**
Clairvaux Creek 143
Clairvaux, Mt VIII, *113*, 131
Claresholm, Alberta 85, *85*
Clark, William 7
Clearwater Forest Reserve *5*, 181, *182*,
    186, 191, 204
Clearwater, Mt *182*, **189**
Clearwater Pass 184
Clearwater River *5*, *182*, **189**
Clitheroe, Mt *113*, 128, **129**, **141**, **178**
Cloudy Ridge 86, **89**
CNoR. *See* Canadian Northern Railway
coal 9, 29–30, 32, 103, 153
Coleman, Alberta 213
Coleman, A.P. (Arthur Philemon)
    42, **70**
Coleman, Mt 198
Coleridge, Samuel Taylor 152

Colin, Mt *113*

Colin Range (mtns) *113*, 122, **180**, **220**, 234, **248–49**, **250–51**

Collie, Mt *57*, 76

Collie, Norman 95

Columbia River and Valley 4, *5*, 6, *7–8*, *56*, 83, 106, 130, 174, 176, 190, 192, 204

Connaught Drive (Jasper) 179, 215

Consolation Lakes *57*, 79

Cook, James 3

Cordillera 29, 66, 69, 211

Coronach Creek *113*, 137

Corps of Royal Engineers 26

Corps of Topographical Engineers (USA) 8

Cortés, Hernán 2

Cory, W.W. 193

Côté, Jean Leon 31, 34

Cottonwood Creek 153

Cougar Brook *56*, 61

Cox, Ross 173, 175

Craigellachie, British Columbia 23

Crandell Lake *86*, **89**, 89

Crandell, Mt *86*, **89**, 89

Crazy Creek *56*, 194

Crowsnest Forest Reserve *5*, 84, *85*, 86–88, **89**, 106, 133, 204

Crowsnest Pass 32, *85*, 88–89

Crowsnest River (formerly West Oldman River) 214

Cruden, David 235

Culture, Ecology and Restoration Project (forerunner of RMRPP) 214–16, 240

Cuthbertson, William 133, 136, 138

Daguerre, Louis Jacques Mandé 25

Daisy Creek *85*, *85*

Dawson, George Mercer 66

Dawson, Mt *56*, 62–63

Dawson, Simon James 10

de Alarcón, Hernando 2

d'Entrecasteaux, Joseph Antoine Bruni 25

de la Vérendrye, Pierre Gaultier de Varennes et 3

de Smet, Pierre-Jean 115

Dennis, John Stoughton 11–13, 46

Department of the Interior 14, 22–23, 53, 64, 66, 72, 79, 83, 102, 138–39, 170, 202, 273

Dominion Parks Branch 36, 102, 106

A Sprig of Mountain Heather (1914) 104, **105**, 152

Forestry Branch 102, 148, 186

Topographical Surveys Branch 12, 22, 24, 28, 139, 198

General Instructions to Surveyors in charge of Parties . . . (1913) 97

Manual of Survey (1884) 13

Topographical Surveys Branch, Calgary Office 192–93, 195

Deville, Daniel-Édouard-Gaston 14, 22, 27–28, 36–37, 58, 106, 118, 132–33, 138–40, 143, 170, 176, 181, 192, 198, 207, 212, 272

appointed Companion of the Imperial Service Order (1916) 140

appointed Surveyor General (1885) 27

death (1924) 194

Photographic Surveying (1889) 28–29, 192

Dixon, Jeremiah 3

Dobbs, Arthur 3

Dominion Lands Branch 14

Dominion Lands Office 12, 196

Dominion Lands Survey (DLS) 12–14, 21, 27, 34, 36, 85, 99, 100, 106, 111, 198–99, 212, 234

Dominion Parks Branch. *See* Department of the Interior

Donkin Pass *56*, 63

Dormer River 182, *182*

Douglas, Howard 102–03

Douglas, Robert 169, 176, 181, 207

Dowling, Donaldson Bogart (D.B.) 30, 34–35, 142

Doyle, Lady Jean 103, 111
Doyle, Sir Arthur Conan 103–05,
 170–72
Drewry, William Stewart 27
Drumheller, Alberta 194
Dungeon Peak (The Ramparts) 166
Dunn, A. 76
Dutch Creek 85, 85

Edith Cavell, Mt. See La Montagne
Edith, Lake 115, 116
Edmonton (and Fort Edmonton),
 Alberta 7, 14, 21, 34, 101, 103,
 107, 109, 132, 138, 149, 194
Edmonton Tent and Mattress Company
 111
Edna Lake 38
Elysium, Mt 113, 131
Emerald Lake 73
Emigrants Mtn 113, 131
Emir, Mt 113, 121
Erebus, Mt 113, 131, 143, 246
Eremite Glacier 246
Eremite Mtn 113, 247
Esplanade, Mt 113, 137, 262–63
Estella, Mt 113, 129
Evans, Lewis 2
Everest, Mt 65, 140, 200
Excelsior, Mt 113, 122
Exshaw, Alberta 156

Fairbank. See Bridgland family farm;
 Parsons family farm
Fairbank United Church (Toronto) 202
Fatigue Mtn 209
Feuz, Edouard (Edward) Jr. 73, 74–75,
 96, 200
Feuz, Ernest, 200
Feuz, Gottfried 73, 74–75, 96
Feuz, Walter 200
Fiddle Peak 113, 220
Fiddle Range (mtns) 135
Fiddle River 35, 100, 103, 113, 133–35
Fiddle River Canyon 135
Fidler, Peter 4
Field, British Columbia 27–28, 57, 73

Field, Mt 57, 66, 76
Finley, Isidore 102, 233–35
Finley, Paulette 20
First World War. See World War, First
Fitzhugh (Jasper), Alberta 36, 101 113
Fitzhugh, Mt 130, 174. See La Montagne
Fitzhugh, Mt (Mt Tekarra) 113, 121
Fitzsimmons, Hugh 96
Fitzsimmons, Mr. 96
Flat Creek 56, 60, 64
Fleming, Sir Sandford 16–18, 21–22, 34,
 39, 91, 215
Folding Mtn 150
Forster, Henry 63
Fort Assiniboine 7
Fort Edmonton. See Edmonton
Fort Garry (Winnipeg) 10
Forty Mile Creek 209
Forum, The (mtn) 113, 143
Frank, Alberta 85, 213–14
Frank Slide 213–14
Fraser, Mt 113, 131, 224–25
Fraser River 5, 16, 106
Front Range (Rocky Mountains) 28,
 92, 150, 156

Geikie, Mt 113, 168, 178
Geikie, Mt 130, 174. See La Montagne
Geodetic Survey of Canada XIII, 235
geodetic surveying 29
Geographic Board of Canada 169
Geographical Names Board of Canada
 92
Geological Survey of Canada 8, 13–14,
 66, 95, 198
Ghost (Angel) Glacier, Mt Edith Cavell
 173, 244–45. See La Montagne
Ghost River 156
Gilmour, A.J. 174
Glacier Park 5, 56, 59, 204, 206
glass plate negatives XIII, 26, 28, 49, 58,
 62, 65, 86, 97, 99–100, 107–09,
 108, 112, 114–15, 118–19, 123–25,
 127–28, 132, 139, 141, 154,
 182–83, 185, 194, 197, 232–33
Gleichen, Alberta 194

Gold Creek 96

Golden, British Columbia 54, 56, 67, 82, 85–87, 205

Gordon, A.M. 76

Gordon, Charles William (Ralph Connor) 200

Grand Trunk Pacific Railway (GTPR) 21, **31**, 32, **33–34**, 39, 91, 94, 101–02, 107, 109, **109**, 111, 136, 138, 150–51, 175, **248–49**, **260–61**

    GTPR and CNoR lines in Jasper Park compared 151

    GTPR stations between Hinton and Jasper 151

    history of completion of 149

Grande Traverse, La Montagne de la. *See* La Montagne

Grant, George Monro 17, **19**, 174–75

Grassie, Charles Andrew 31

Gray, Robert 7

Great Britain

    The British North America Act (1867) 196

Green, Ashdown 20

Green, H.J. **109**

Group of Seven 115, 147, 154

GTPR. *See* Grand Trunk Pacific Railway

Harkin, James Bernard 102, 104, 106, 150, 170, 176, 181, 239

Harmon, Byron 200

Harris, Ley Edwards 60, 181, 184–87, 190–92, 194, 213

Harris, Mt *182*, 184–86, **189**

Hasler, Christian Jr or Sr 84

Hatch, Helen 78

Hawkins, Albert Howard 29, 136, 139, 155

Hector, Sir James 9, 30, 95, 143–44, 174

Henry House Flats *113*, **180**, **250–51**

Henry, Mt *113*, 131

Herdman, Rev. J.C. **70**, 73

Hermit, Mt 56, 78

Herriot, George Henry 35–36, 136, 155

Hess, Frederick 68, 95, 201

Hess, Ina 68, 80, 95, 201

Hess, Margaret B. (Marmie) IX–X, XIII, 81, 95, 200–201, 236, 272

Hess Special Collections Library (University of Calgary), Margaret B. 209

Higgs, Eric **121**, **180**, 214–20, **217**, 222–24, **225**, 226–28, **227**, 231–33, 235–36, **246–47**, 249, **251**, **253**, **255**, **257**, **259**, **261**, **263**, 274

Hill, James Jerome 21

Hime, Humphrey Lloyd 26

Hind, Henry Youle 10, 26

Hinton, Alberta 109

Holway, E.W.D. 174

Horetzky, Charles George **18–19**, 39–40

Horseshoe Lake (Lac Beauvert) 111, *113*, **116**, *123*, 153, 209, 218, **252–53**

Howard, G.E. 174

Howse Pass 19, 56, 67, 95

Huber, Emil 63

Huber, Mt 57, 79

Hudson's Bay Company (HBC) 2, 4, 7, 9, 11–12, 80

Hunter, Edwin 41

Hyatt, Alfred Edward 85, 88–90, 97, 107, 112, 114–15, 118, 121, 123–27, 131–34, 138, 144, 149, 154–55, 217–18

    death of 131–34

iconometry. *See* photogrammetry

Illecillewaet Glacier 62

Illecillewaet River 56, 59

Imperial International Boundary Commission 14, 26

Incomappleux River 60, 64

Indian Ridge *113*, 131

Interlaken 138. *See also* Grand Trunk Pacific Railway: GTPR stations between Hinton and Jasper

Jackson, A.Y. 115, 147

Jacques Creek 138

Jasper, Alberta 36, 101, 110, *113*, 114, **117**,
          122, 127–28, 133, 136, 138, 172,
          174–75, 195, **252–53**, **256–57**,
          258–59
Jasper East Mtn. *See* Signal Mtn
Jasper Forest Park 102, 150
Jasper House I 9, *113*
Jasper House II 9, **18–19**, 19–21, 29, *113*
Jasper Lake 20, 38, *113*, 133, **137**, 174, 213
Jasper National Park (after 1930) XII,
          200, 206, 215, 228–29, 234, 274
Jasper National Park Library 233
Jasper Park (Jasper National Park from
          1930) XII, 4, *5*, 29, 35–36,
          99–138, *113*, 151, *182*, 195, 204,
          208
     boundaries of redrawn 150, 156
     ecological change in during the
          twentieth century 229–31
     history of the establishment of
          101–03
     removal of Métis homesteaders
          from 102, 179, 233–35
     survey of by Bridgland 99–138
Jasper Park Collieries 31, 42
Jasper Park Lodge. *See* Tent City
Jasper Park Lodge golf course 154, 207,
          258
Jasper-Yellowhead Museum and
          Archives 233
Jefferson, Thomas 15
Joachim, Adam 102

Kain, Conrad 79
Kananaskis Range (mtns) 156
Kataka, Mt *113*, 145
Kenneth, Robert 111
Kicking Horse River 57, 101
King, William Frederick 14
Kipling, Rudyard 52
Kittson, Norman 21
Klondike Gold Rush 23, 52
Klotz, Otto Julius 14, 29, 66
Kootenay Park 5, 57, 65–66, 187, 190–91,
          204
Kootenay River 190, 192

Lake Louise, Alberta 41, 57, 66, 77, 79,
          109, 195
Lambert, Johann Heinrich 25
La Montagne de la Grande Traverse
          (Mt Edith Cavell) *113*, 125, 130,
          172–76, **177**, 179, **244–45**
Lamoureux, Alfred 30
Land Ordnance (USA) 4
Larmour, Judy 36, 273
Laussedat, Aimé 25–27
Le Duc, Mt (La Montagne) 174
Lewis, Merriwether 7
Library and Archives Canada (National
          Archives) XIII, 202, 218, 235–36
Livingstone Range (mtns) 85, *85*
Livingstone River 85, *85*
Logan, Sir William Edmond 8–9
Loudon Creek 92
Loudon, James 92
Loudon, Mt 92, *182*
Loudon, Thomas Richardson 92, 209
Loudon, William James 51, 54, 92, 209
Lower Waterton Lake 86
Lundbreck, Alberta 85, 87, 112

McArthur, James Joseph 23, 27, 29, 100,
          148, 206, 211
McArthur, Lake 57, 201
McConnell, George 66
McCrae, A.O. 76
Macdonald, Sir John A. 12, 22
McDougall, William 11–12
McEvoy, James 30
McFarlane, J.B. 136, 139, 155
McGillivray, William 5
McIntyre, Duncan 21
Mackenzie, Alexander 7
McKinnon, A. 132
McLaggan, John W. 102, 153
MacLaren, I.S. 235, **261**, 272, 274
MacLeod, Alec M. 153, 236
MacLeod, Henry A.F. 34, 42
McNicoll, David 72
McTavish, P.D. 73
Maccarib Creek 128

Maccarib, Mt *113*, 127–28, 130, 144, 176, **177**, 222

Maccarib Pass 128

Macoun, John 22

Main Range (Rocky Mountains) 27, 65, 67, 110, 148

Majestic, Mt *113*, 128, **129**, **204**

Makwa Ridge *113*, 145

Maligne Gorge/Canyon **114**, 115, 122

Maligne Lake/Sorefoot Lake 34, 55, 132

Maligne Range (mtns) 138

Maligne River 110, *113*, 114, 132–33

Mallory, Sir George Leigh 200

Mansfield, Jared 15

Manx Mtn *113*, 126

map-making 140–45

Marmot Mtn *113*, 126, 131

Marmot Pass 126

Marpole, Mt *57*, 76

Martin Creek **189**

Martin, J.W. 122, 127–28, 131, 138 identity of discussed 155

Mason, Charles 4

Matheson, Hugh 36–37, 111, 139, 142, 155

Meadow Creek *113*, 128, 143

Medicine Lake *113*, 132, **254–55**

meridians (grid survey) 13–15, 29, 38, 136, 139

Mesa Verde National Park (USA) 201

Michel, Friedrich 63

Middle Waterton Lake 86

Miette 136. *See also* Canadian Northern Railway: CNoR stations between Hinton and Jasper

Miette Hot Springs **35**, 35, 103, *113*, 134–35, 175, **227**

Miette River and Valley 16, 19, 99, 110, *113*, 116, **117**, 125, 128, 131, 174

Minnewanka, Lake 28, *57*, 88, 209

Missouri River 7

Mitchell, S.H. (Stanley) **70**

Moat Lake **168**, **178**

Moberly, Adolphus 102

Moberly, Ewan 102

Moberly, John 102

Moberly, Walter 18–21

Moberly, William 102

Monashees (mtns) *56*, 84

Moosehorn Creek *113*, 137

Morley, Alberta *57*, 80, 87, 184

Mount Revelstoke Park *5*, *56*, 206

Mount Robson Provincial Park 116

Mountain Creek 153

Muhigan, Mt *113*, 131, 145

Mumm, A.L. 174

Murchison Group (mtns) 92

National Air Photo Library (Ottawa) 208

Natural Resources Canada, Legal Surveys Division XIII

Nelson Mountains (Selkirks) 6

Nicholson, Claus 187

Niépce, Joseph Nicéphore 25

Nigel Creek *182*, 195

Nipika Mountain Resort 190

Norfolk, E. 123, 126, 132–33, 138, 155, 217

North Saskatchewan River *5*, 6, 92

North West Company 4

North-West Rebellion 23, 93

Northern Railway 47–48, 91

Ochre Ridge 86, 89

Odaray, Mt *57*, 79

Ogilvie Range (mtns) 23

O'Hara, Lake *57*, 77, 79

Oil City, Alberta 86, **88–89**

Okotoks, Alberta 85

Old Bow Fort 80–81

Old Fort Creek 80

Old Fort Point **122**, 122, **252–53**

Old Peigan Post 80–81

Oldhorn, Mt *113*, 130

Oldman River *5*, 85–86, *85*

Oliver, Mt *56*, 64

Ontario SPS (School of Practical Sciences), University of Toronto 50, 91

Oregon Treaty 10

Ottawa, Ontario 139, 145, 183

Otterhead River *57*, 58

Otto Brothers (Bruce, Closson, Jack) 138

Otto, Jack 70, 75, 138
Outram, Sir James 64

Palisade, The *113*, **234**
Palisades Centre *113*, 226
Palliser, John 9, 174
Panther River 182, *182*
Paradise Valley *57*, 77
Parker, Elizabeth 69, **74**, 176
Parsons, Annie Jane (*née* Bridgland) 48
Parsons, Clarke 48
Parsons, Elizabeth Sophie (*née* MacKay)
        48
Parsons family farm (Downsview,
        Ontario) **47**, 48–49, 91, 201
Parsons, Jacob 45, 48
Parsons, Matthew 48
Parsons, William Abbott 48
Patricia Lake 115
Patricia, Mt 133–34
Patterson, J.D. 79
Peace River District 195
Pearce, William 23
Pembina, North Dakota 11
Pennsylvania 32
Perkins, Charles 67
Perkins, George 67
Perkins, Ina 67
Peters, Frederic Hatheway 193–94,
        197–99
photogrammetry 25, 27, 29, 100, 140, 196
photography, aerial 100, 192, 196
phototopographical surveying,
        comparative costs of 41.
        See also surveying
Pincher Creek, Alberta 15
Pipestone Pass *57*, 66
Pitt Lake 186–87
Pocahontas, Alberta 31, **32–34**, 35, *103*,
        110, *113*, 133–36, **134**, 175
Pollard, Harry 200
Porcupine Hills, Alberta 15
Portal Creek *113*, 126–28, 131, 175
Portal Peak 143
Postern Mtn **168**, **178**
Prairie de la Vache *113*, 173

Prime Meridian Conference (1884) 16
Prince Rupert, British Columbia 149
Purcell Mtns 54, *56*
Purity, Mt *56*, 187
Pyramid Lake 36, 115, 153
Pyramid Mtn *113*, 115, 123, **252–53**

Racehorse Creek 85, *85*
Railway Belt (across British Columbia
        from Field west along CPR)
        27–28, 82–83, 87, 97, 106, 110
railway construction 10, 12, 29–30.
        See also Canadian Northern
        Railway; Canadian Pacific
        Railway; *and* Grand Trunk
        Pacific Railway
Ram River 186
Ramparts, The *113*, 128, **129**, **141**, **168**,
        178, **204**
Red Deer River 5, 6, 182, *182*, 184
Red River Settlement 11
Redoubt Peak **166**
Reid, Dennis 147
rephotography (repeat photography)
        XII, XIII, XIV, 216, **220**, 223,
        226, 232, 236, 241
    analysis of 231–32
    compared to Bridgland's survey
        photographs 228–31, 243–63
    digitalization in 231, 233–35
Revelstoke, British Columbia 29, 54, *56*,
        59, 83–84, 132, 192, 194–95,
        204–05, 218
Rhemtulla, Jeanine **178**, **180**, 214–23, **217**,
        220, **225**, 226, **227**, 229–32,
        235–36, **246–47**, 249, 251, **253**,
        255, **257**, **259**, **261**, **263**, 273–74
Riding Mountain National Park 197
Riel, Louis 12, 23, 93
Riel Rebellion (Red River Rebellion) 12
Riley, T. 27, 41
RMRPP. *See* Rocky Mountain Repeat
        Photography Project
Roberts, Charles G.D. 52
Robertson, J. 132
Robson, Joan 195

Roche à Bosche *113*, **221**
Roche à Perdrix 110, *113*, **135**
Roche Bonhomme *113*, 122, **220**
Roche Miette 8–9, 17, **18**, 19, 21–22,
    29–30, **31**, **109**, 110, *113*, 134,
    134–35, 153, **260–61**
Roche Noir *113*, 143
Roche Ronde *113*, 226
Rocky Mountain Fort. *See* Jasper House I
Rocky Mountain Repeat Photography
    Project (RMRPP) XII, XIII,
    XV, XVII, 154, 214–39, **238**,
    241, 243–63, 274. *See also*
    rephotography
    The Rockies through the Lens of Time
      (2003) 238–39, 274
    Vantage Points: Scientific Photography in
      Jasper National Park (2005–06)
      238–39, 274
Rocky Mountains 100, 237
Rocky Mountains (Banff) Park *5*, 28,
    57, 64–65, 80, 92, **105**, 111, 156,
    181, *182*, **189**, 195, 207
Rocky River 21, 133, 138
Rogers, Mt *56*, 78
Rogers Pass *56*, 60–61, **62**, 63, 78, 80
Rogers Pass, British Columbia 54, *56*, 59
Rogers, Samuel Maynard 36, 102, 152, 209
Rostrum, The (mtn) *113*, 143
Ruby Ridge 86, **88–89**, 89
Rummell, Lizzie 181, 273
Russell, Alexander 14
Russell, Lindsay 13–14, 22
Ryerson, Adolphus Egerton 46

St Cyr, Arthur 29, 136, 139, 155
St Elias Range (mtns) 23
St Hilda's School for Girls (Calgary) 68
St Paul, Minneapolis and Manitoba
    Railway 38
St Paul, Minnesota 10
Salmon Arm, British Columbia 84
Sarcee (Tsuu T'ina) 101
Saskatchewan Glacier 195
Saskatchewan River 4, 106

Sawback Range *57*, 65
Schäffer, Mary 34–35, 55
Scott, Thomas 12
Second World War. *See* World War,
    Second
Seebe, Alberta *57*, 80
Selkirk Mtns 30, 54–55, *56*, 58, 78, 84,
    100, 110, 148, 187, 200, 205, 212
Service, Robert 52–53
Seton, Ernest 52
Shaw, Adrienne 236
Sheol, Mt *57*, 65
Sherbrooke, Lake *57*, 79
Shuswap Lake 82, 192
Siffleur Mtn 92
Siffleur River *182*, 184
Siffleur Wilderness Area 92, 186
Signal Mtn *113*, 114–15, **117**, 118, 121–22,
    154, **258–59**
Simpson, Æmilius 7
Simpson Pass **105**, 152
Simpson, Sir George 6
Sissons, Charles Bruce 51, 58, 72–73, 75,
    83, 94, 197, 204, 206, 272
Sissons, Frank 206
Small, Charlotte 6
Smith, Donald Alexander (Lord
    Strathcona and Mount Royal)
    21
Smith, Frank 32
Smith, Trudi **108**, 236, **238**, 274
Smithers, British Columbia 32
Smithsonian Institution 66, 205
Snake Indian River *113*, 137
Snaring River *113*, 137, 215, **216–17**,
    229–31, **234**, **262–63**
Spray Lakes *57*, 88
SPS. *See* Ontario SPS
Stephen, George 21
Stephen, Mt 27, *57*, 66
Storm Mtn *57*, 84
Stutfield, Hugh 95
Sulphur Creek **227**
Surprise Point *113*, 130, **166**, 223, **224–25**

surveying 29
  advent of photopographic alpine
      surveying in Canada 27
  Canadian and US American systems
      of grid survey compared 13
  changes in elevation a challenge to
      chain grid township surveying
      24–25
  CPR Survey. *See* Fleming,
      Sir Sandford
  first use of photography in North
      American surveys 26
  grid township system approved for
      prairie survey 13. *See also*
      meridians; *and* base lines
  and railway construction 16–22
  and Red River Settlement 11
  repeat phototopographic surveying
      as compared to Bridgland's
      222–28
  stadia surveying 42–43, 155, 193
  triangulation 40–41
  triangulation surveys and
      photography 26
Surveys Branch. *See* Department of
      the Interior, Topographical
      Surveys Branch
Swift, Lewis 102, 150
Syncline Ridge (Roche Miette) 19, **109**

Taft, British Columbia 56, 194
Talbot Lake 38
Tekarra (Fitzhugh), Mt *113*, 121, 143
Tekarra (Iroquois guide) 10, 143–44
Temple, Mt *57*, 77
Tent City 111, 115, 123, 153
Tête Jaune Cache, British Columbia
      103, 149
theodolite. *See* transit telescope
Thompson, David XI, 4–7, 6, 67, 101, 206
Thomson, Tom 146–47
Thumb (Signal Mtn) *113*, 154
Thunderbolt Peak *113*, 130, **246–47**
Tonquin Hill 130, **168, 178**
Tonquin Pass **168, 178**

Tonquin Valley 100, *113*, 125–26, 128, **129**,
      130–31, **166**, 200, 223
Toot Toot (Mt Bridgland) VIII, *113*
Topham, Harold Ward 63
Topley, Horatio Nelson 28
Topley, William James **110, 114**
Toronto Junction Collegiate Institute
      (Humberside Collegiate) 49
Toronto, Ontario **47**, 199, 202
transit telescope 13–14, 26–27, 42, 99,
      112, 115, 118, 120, 122, 127, 130,
      133, 136, 138, 142, 184, 194–95, 233
Troughton and Simms 115
Turner Valley, Alberta *85*, 205
Turnor, Philip 4
Turret Mtn **168, 178**
Turtle Mtn *85*, 214

Union Internationale des Associations
      d'Alpinisme (UIAA) 226
United States Geological Survey 14–15
University of Toronto 92. *See also*
      Victoria College/University
      (Toronto); *and* Ontario SPS
Upper Hot Springs (Banff) 78
Upper Waterton Lake 86
Ursus Major, Mt *56*, 61
Ursus Minor, Mt *56*, 61
Utopia, Mt *113*, 136, 226–28, **227**

Valad 174
Vancouver, British Columbia 199
Vancouver, George 7
Vaux, Mary 41
Verdant Valley **177**
Vérendrye, Pierre Gaultier de Varennes
      et de la 3
Vertex, Mt *113*, **129**, 131
Vicary Creek *85*, *85*
Vice President, The (mtn) *57*, 72–73, **76**
Victoria and Albert Museum (UK) 202
Victoria, British Columbia 199
Victoria College/University (Toronto)
      46, 49, **51**, 91
  Acta Victoriana 49, 272

Victoria Cross Range (mtns) 206
Villeneuve, Frank 30

Wabasso Lake *113*, 145
Walcott, Charles Doolittle 66, 205
Wallace, Rod 215
Wapta, Mt *57*, 66, 76
Warren, Gouverneur Kemble 7
Washington, George 3
Waterton Lakes Forest Park 84
Waterton Lakes Park *5*, 15, 84, 85–86,
    88–89, 90, 106
Waterton Lakes National Park XIII, 236
Waterton River 24, 86
Watt, Rob 236
Webb, Jenaya 146, 233
Weldwood of Canada (Hinton) 219
Wheeler, Arthur Oliver XI, 23–25,
    29–30, 41, 54, 58–65, **62**, **70**,
    72, **74**, 77–79, 84, 87, 93–94,
    100, 140, 148, 156, 176, 181, 193,
    199, 211–12, 272
  character 55, 94
  The Selkirk Mountains (1912) 176
  The Selkirk Range (1905) 59, 176
Wheeler, Clara **74**
Wheeler, Edward Oliver 65, **74–75**, 81,
    200, 206, 272
Wheeler, Hector 58, 60, 63, 65, 72–73
Wheeler, Oliver 58, 64
Whirlpool River 10, *113*, 125, 130,
    173–74, 176, **177**

Whistlers Creek *113*, 126, 175
Whistlers, The (mtn) *113*, 126, 175, 223,
    256–57
Whymper, Edward 72
Whyte Museum of the Canadian
    Rockies (Banff) 214
Wildland Recreational Area 186
Williams, Mabel Bertha 181
Willingdon, Mt *182*, 184, **189**
Wilson, L.C. 77, 200
Wilson, Mt *182*
Wilson, Tom 41, **70**
Winderemere Road 190
Winnipeg, Manitoba 32, 70
Wood River *113*, 130
World War, First 73, 94, 111, 205, 209, 214
World War, Second 100
Wratten and Wainwright 108–09

Ya-Ha-Tinda (Brewster) Ranch *182*, 208
Yearling, Bill 190
Yellowhead Highway 150, 153
Yellowhead Pass 16, 19, 22, 39, 94, 104,
    116, 138, 149
Yoho Park *5*, *57*, 65, 67, 79, 205
Yoho Pass 73
Yoho River and Valley *57*, 72, 94

Zezulka-Mailloux, Gabrielle 234, 272,
    274